Macmillan Building and Surveying Series

Series Editor: Ivor H. Seeley
Emeritus Professor, Nottingham Trent University

Accounting and Finance for Building and Surveying A. R. Jennings
Advanced Building Measurement, second edition Ivor H. Seeley
Advanced Valuation Diane Butler and David Richmond
Applied Valuation, second edition Diane Butler
Asset Valuation Michael Rayner
Building Economics, fourth edition Ivor H. Seeley
Building Maintenance, second edition Ivor H. Seeley
Building Maintenance Technology Lee How Son and George C. S. Yuen
Building Procurement Alan E. Turner
Building Project Appraisal Keith Hutchinson
Building Quantities Explained, fourth edition Ivor H. Seeley
Building Surveys, Reports and Dilapidations Ivor H.Seeley
Building Technology, fifth edition Ivor H. Seeley
Civil Engineering Contract Administration and Control, second edition Ivor H. Seeley
Civil Engineering Quantities, fifth edition Ivor H. Seeley
Civil Engineering Specification, second edition Ivor H. Seeley
Commercial Lease Renewals – A Practical Guide Philip Freedman and Eric F. Shapiro
Computers and Quantity Surveyors A. J. Smith
Conflicts in Construction – Avoiding, Managing, Resolving Jeff Whitfield
Constructability in Building and Engineering Projects Alan Griffith and Tony Sidwell
Construction Contract Claims Reg Thomas
Construction Estimating, Tendering and Bidding A. J. Smith
Construction Law Michael F. James
Contract Planning and Contract Procedures, third edition B. Cooke
Contract Planning Case Studies B. Cooke
Cost Estimation of Structures in Commercial Buildings Surinder Singh
Design-Build Explained D. E. L. Janssens
Development Site Evaluation N. P. Taylor
Environmental Management in Construction Alan Griffith
Environmental Science in Building, third edition R. McMullan
European Construction – Procedures and Techniques B. Cooke and G. Walker
Facilities Management – An Explanation Alan Park
Greener Buildings – Environmental Impact of Property Stuart Johnson
Housing Associations Helen Cope
Housing Management – Changing Practice Christine Davies (Editor)
Information and Technology Applications in Commercial Property Rosemary Feenan and Tim Dixon (Editors)
Introduction to Building Services, second edition Christopher A. Howard and Eric F. Curd
Introduction to Valuation, third edition D. Richmond
Marketing and Property People Owen Bevan
Principles of Property Investment and Pricing, second edition W. D. Fraser
Project Management and Control David Day

Property Finance David Isaac
Property Valuation Techniques David Isaac and Terry Steley
Public Works Engineering Ivor H. Seeley
Quality Assurance in Building Alan Griffith
Quantity Surveying Practice Ivor H. Seeley
Recreation Planning and Development Neil Ravenscroft
Resource Management for Construction Mike Canter
Small Building Works Management Alan Griffith
Structural Detailing, second edition P. Newton
Sub-Contracting under the JCT Standard Forms of Building Contract Jennie Price
Urban Land Economics and Public Policy, fifth edition Paul N. Balchin, Gregory H. Bull and Jeffrey L. Kieve
Urban Renewal – Theory and Practice Chris Couch
1980 JCT Standard Form of Building Contract, second edition R. F. Fellows
Value Management in Construction Brian R. Norton and William C. McElligott

Accounting and Finance for Building and Surveying

A. R. Jennings

B.A. (Hons) F.C.C.A.

First published 1995 by
MACMILLAN PRESS LTD
Houndmills, Basingstoke, Hampshire RG21 2XS
and London
Companies and representatives
throughout the world

ISBN 0–333–60961–1

A catalogue record for this book is available
from the British Library.

10 9 8 7 6 5 4 3 2 1
04 03 02 01 00 99 98 97 96 95

Copy-edited and typeset by Povey–Edmondson
Okehampton and Rochdale, England

Printed in Great Britain by
Antony Rowe Ltd, Chippenham, Wiltshire

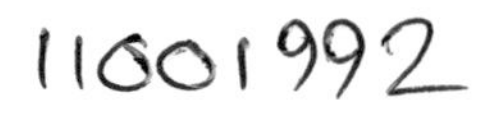

Contents

Preface

Accounting is a topic encountered in all sizes of businesses and in all business sectors. Relatively few non-accountants understand what accounting aims to do or what the figures mean when presented. This is in some measure a result of the specialised terminology used in accounting, which imparts precision and avoids ambiguity. Because of this, accounting and finance are in many people's minds surrounded by an impenetrable mystique.

The purpose of this book is to explain, as directly and lucidly as possible, accounting terminology, concepts, practices and procedures, relating these wherever possible to building and surveying. This should enable readers to communicate more effectively with accountants, to put accounting information and data to far better use and to be in a position to ask accountants for information, the existence and significance of which non-specialists were previously unaware.

Very broadly, the two roles of accountants covered by this book are financial scorekeeping and financial navigation. The first of these roles is dealt with in Part I – Financial Reporting and Analysis, and the second in Part II – Management Planning and Control Practices.

The needs of student members are further met by the inclusion of an appendix (Appendix III) containing a selection of typical questions of an accounting nature taken from external examination papers. These are accompanied by full solutions cross-referenced to chapters and sections within this book.

Nottingham A. R. JENNINGS

Acknowledgements

The author wishes to record his thanks to the Board of Directors, John Maunders Group plc, for permission to reproduce extensive extracts from the Annual Report and Accounts, and to the Chartered Institute of Building for permission to use selected questions from Member Examination Part II examination papers.

PART I

Financial Reporting and Analysis

Part I of this book deals with aspects of the preparation and presentation of financial information of businesses which affect, or are of interest to, outside individuals and bodies.

In contrast to this part, which is concerned with external reporting matters, Part II deals with the financial aspects of planning and controlling the operations of a business from an internal management perspective.

1 Types of Business Formation

1.1 Introduction

The form of structure of a business is largely dependent upon the size and complexity of the operations on which it is engaged. This chapter examines business organisational structures from the very simple to the very complicated.

A basic distinction between companies is that they are either *incorporated* or *unincorporated*:

- An *incorporated business* is one in which the legal identity of the inanimate business is separate from that of its members, thus enabling it to enter into contracts, to sue and generally to conduct business in its own name.
- An *unincorporated business* is one in which the individual members are legally responsible in their private capacities and to the limit of their private resources in the conduct of the business.

A feature of all businesses is that they are formed with a view to making a profit.

Examples of each of these types of business now follow, in ascending order of complexity.

1.2 Unincorporated businesses

The two common forms are *sole proprietorship* and *partnership*:

Sole proprietorship is the simplest business formation in a start-up situation when an individual decides to become self-employed. There are numerous examples in everyday life of self-employed craft operatives, plumbers, plasterers, joiners or electricians, for example, and for professionals, including surveyors, architects, accountants and solicitors.

The business of a sole proprietor can be conducted either in the name of the proprietor (owner) or in a specially devised business name. For example, a self-employed plumber, J. Wilson, could trade in his own name or choose a separate name, say, Capital Plumbing. In either case, Wilson himself would be fully responsible for settling all the debts of his business. Apart from this legal obligation, Wilson both owns and manages the business. This does not mean that he has to work alone. He could engage employees to cope with a heavy workload but they would not receive the rewards or bear the responsibilities of ownership; if they were to have any involvement in management, as a foreman, for example, it would only be by delegation from Wilson, the owner, who would remain ultimately responsible for the acts or omissions of his employees.

Partnership is the term given to a business consisting of several joint owners. Obviously the minimum number of owners is two. For the majority of businesses the upper limit is fixed by law at twenty, above which figure it would be an illegal association. The exceptions are that a banking partnership is restricted to a maximum of ten partners, but for accountants, solicitors and stockbrokers, it is fifty.

A partnership can be formed as such from the outset but it often evolves from a sole proprietorship or by the amalgamation of two such businesses. Typical examples of partnerships can be found among surveyors, architects and estate agents. A business can engage in a much larger scale of operations as a partnership than as a sole proprietorship. This is because all the individual partners contribute physical resources, such as premises, or intangible resources including business or professional expertise, or useful business or customer/client contacts.

Formation of a partnership is by means of an agreement which can be verbal but, in the interests of the partners, it should be in writing to minimise the possibility of disputes at a later stage over what was intended. A typical partnership agreement would deal with such matters as the proportions in which partners would share profits and losses from trading activities and, in the event of the partnership being dissolved, the respective monetary contributions of the partners, entitlement to salaries of individual partners, entitlement to interest on monetary contributions and loans by individual partners, and liability of individual partners to pay interest on interim withdrawals of profits.

As regards the partnership name, a somewhat similar situation arises as with a sole trader. The business can be conducted in the names of the partners or in a business name. Thus, if the self-employed plumber, J. Wilson, formed a partnership with A. Vickers and G. Yorke, the business name could be Wilson, Vickers and Yorke, or Wilson and Partners, or, as previously, Capital Plumbing. The partners are all liable, to the full extent of their private resources, for settling partnership debts.

1.3 Incorporated businesses

This label embraces what are known as *limited companies*. A company has a legal existence separate from that of its members. The legal liability of the members of a company is limited to the sums they have contributed to become members, plus any further sums they have agreed to contribute on a future occasion or occasions. Unlike the situation of a sole proprietorship and a partnership, the members of a limited company are not responsible for settling any debts of the business. The business itself is responsible and if it is unable to discharge its debts, it can be declared insolvent (the corporate equivalent to being bankrupt, in the case of a person) and compulsorily wound up – that is, forced out of existence by a process of liquidation whereby its possessions are sold off and the proceeds used to pay off some of the debts.

A minimum of two members are needed when a limited company is formed; there is no upper limit, except for a private company. Some large companies have thousands of members (who may be private individuals or corporate entities) known as *shareholders*. It is these shareholders who are the owners. The management of the company in the day-to-day affairs is, at the highest level, in the hands of officers employed for the purpose and designated as *directors*, acting together as a Board of Directors, under a Chairman. This arrangement whereby the ownership and management of the business is in entirely separate hands distinguishes the organisational structure of a limited company from that of a sole proprietorship and of a partnership. Initially, directors are voted into office by the shareholders in a general meeting. The usual arrangement is for a proportion of the directors to retire annually by rotation and to offer themselves for re-election, if they so wish. Difficulties which would ensue if the whole Board were to retire simultaneously are thereby avoided.

The earliest stages in the creation of a new company are carried out by formation specialists known as promoters, who subsequently may or may not become shareholders and/or directors. The promoters carry out the legal formalities needed to bring a company into existence. One of the vital documents that the promoters have to file with the Registrar of Companies is the Memorandum of Association of the proposed company. This document sets out various matters affecting the relationship of the company with outside parties. Matters covered include the domicile of the company, the location of the company's registered office, the objectives the business will be entitled to pursue and the funds it will be able to raise by subscription and borrowing. The internal affairs of the company are regulated by another document, the Articles of Association, which deal with such matters as conduct of meetings, shareholders' rights, voting rights, powers, duties, appointment and removal of directors, and various other matters. On completion of the legal formalities the company comes into being on the issue by the Registrar of Companies of a Certificate of Incorporation. This does not entitle the company to start external operations. Further formalities have to be concluded satisfactorily before a Trading Certificate, enabling the company to commence business operations, is issued.

There are two classes of company limited by shares – *public* and *private*:

- A *public company* must register as such and contain the words 'public limited company' as an integral part of its name; this may be abbreviated to PLC or plc, or, in the case of a Welsh company, ccc. It can raise funds from the general public by subscription and borrowing. There are onerous reporting requirements for a plc; these are dealt with in Chapters 3 to 6.
- A *private company* is any company that is not a public company. It must contain the word 'Limited' (or Ltd) as an integral part of its name. A private company is prohibited by the Companies Act 1985 from raising funds from the general public. For this reason the shareholders (owners) are often family members or other connected individuals. Reporting requirements for a private limited company are less onerous than those of a plc, to enable a degree of confidentiality to be exercised over the financial affairs of its members.

1.4 Other business formations

Apart from individual, freestanding companies there are two other common business formations. These are *joint ventures* and *not-for-profit organisations*:

- A *joint venture* is an arrangement, similar in nature to a partnership but without the formal arrangements, between two or more incorporated and/or unincorporated parties. It is entered into for a limited period of time or for the achievement of a defined objective. Although, as in the case of a partnership, the parties contribute financial and non-financial resources and share profits and losses in pre-agreed proportions, they retain their separate identities and are liable individually, at all times; unlike a partnership, a joint venture is not a separate entity.

  Joint venture arrangements have enjoyed increased popularity in recent years. London and Edinburgh Trust has formed a joint venture with Highland Participants for the development of Southampton Airport; and National Power has in at least one instance entered a joint venture with a commercial market gardener, whereby water used to absorb heat from power generation equipment is piped to the commercial grower and utilised to heat greenhouses. A massive international joint venture, worth 3.5 billion Hong Kong dollars, for a landfill project was formed between Waste Management International of London, Citic Pacific Ltd of China, and a Hong Kong Company, Sun Hung Kai Properties Ltd.

  The ten joint venturers in TransManche Link (TML) engaged in building the Channel Tunnel for Eurotunnel comprised Balfour Beatty, Costain, Tarmac, Taylor Woodrow, Wimpey and five French contractors.

- *Not-for-profit organisations,* which may or may not be incorporated, exist primarily to render a service, in the course of which they may make a profit, but this is not their objective. Many of these organisations are unincorporated, typical examples being amateur football, cricket and social clubs, and amateur dramatic, operatic and musical societies. Examples found in the housing sector include sheltered housing schemes, residents' associations and housing associations, many of the latter being incorporated bodies.

1.5 Forms of business combination

There are various ways in which businesses can combine, either informally or formally. Under informal arrangements businesses can co-operate with each other on large projects by forming a *consortium.*

- A *consortium* is an association of businesses whose combined resources are engaged on a project. It is not a legal entity as such, as each business retains its separate identity, nor are profits and losses shared in pre-agreed proportions as they are in the case of a joint venture. Each business therefore maintains its independence and is not a sub-contractor of any other business within the consortium; as a consequence, each business retains any profits made on the project (or bears any losses sustained) and meets only its own liabilities. Large civil engineering contracts/projects are often undertaken by a consortium of contractors, none of which individually has sufficient physical and/or financial resources or expertise to proceed alone.

Formal methods of combination are more numerous, five of which will now be explained – *amalgamation, absorption,* and the *acquisition of a controlling interest by share purchase,* share *exchange* or *contract.*

- An *amalgamation* is the term used when two or more existing businesses are wound up and cease to exist as individuals but their combined resources are acquired by an entirely new company formed for the purpose. Businesses involved in an amalgamation are usually of approximately similar size.
- An *absorption* is said to occur when a relatively large, dominant business acquires the resources of one or more smaller businesses. These absorbed businesses are wound up and cease to exist as individuals but their resources then augment the existing resources of the acquiring business.
- An *acquisition of a controlling interest* differs from the two preceding types of combination in that it involves the acquisition not of resources but of shares. The acquiring business secures a sufficient number of

shares from the shareholders of the targeted companies to give it control by virtue of a majority of voting rights. These rights are then exercised to give the acquirer at least one seat on the Board of Directors of the companies concerned and also to ensure that resolutions voted upon in general meeting can be passed or rejected as required. An alternative to the acquisition of a majority of the voting shares is for the acquirer to gain control by means of a control contract with the targeted companies or by the inclusion by them of provisions in their Memorandum of Association or Articles of Association (see Section 1.3) under which control is given to the acquirer. Irrespective of the means by which control is gained, the acquired companies continue their separate existences and operations, but on policies dictated by the acquirer. Where control is gained by means of shares, the situation is either an acquisition or a merger.

- An *acquisition* takes place if the acquiring company buys the shares for cash from the shareholders of the targeted companies.
- A *merger* is said to have occurred if the acquiring company exchanges its own shares for those, or for not less than 90 per cent, of the equity shares of the targeted companies. Thus, if Z plc has an ordinary share capital of 2 000 000 shares of which A plc acquires not less than 1 800 000 (90 per cent), A plc has a controlling interest in Z plc. If A plc buys these shares for cash, this constitutes an acquisition; but if it is a share exchange situation and no more than 10 per cent of the nominal value of the equity shares issued is in the form of a cash consideration, then this is regarded as a merger. The significance of the distinction between an acquisition and a merger is seen in the presentation of financial figures dealt with in Chapters 3 and 4.

1.6 Groups

Group is the technical term applied to the investing company, that is, the company acquiring a controlling interest, together with not only those undertakings, both incorporated and unincorporated, it is able to control, but also those over which it can exercise a considerable degree of influence. In order to understand this arrangement, a number of technical labels must be understood; these are *parent undertaking, participating interest, subsidiary, sub-subsidiary, associated undertakings and minority interests*:

- The *parent undertaking* (previously known as the holding company) is the company which has acquired varying degrees of control, as described in Section 1.5, over other undertakings (which can be incorporated or unincorporated bodies), termed subsidiary or associated undertakings, according to circumstances.

- A *participating interest* is an investment by a parent company as a long-term venture and for the purpose of securing a contribution to its activities by the exercise of control or influence. The investment may consist of a shareholding or an equivalent interest; a holding of 20 per cent or more of the shares of an undertaking is deemed to be a participating interest unless this presumption can be rebutted, for example, if another investing company can exercise a greater degree of influence.
- A *subsidiary undertaking* is one in which a parent undertaking holds a participating interest and either the two are managed on a unified basis, or the parent actually exercises a dominant influence. For example, Haven Retirement Homes Ltd and Mancunian Developments Ltd are both subsidiary undertakings of John Maunders Group plc.
- A *sub-subsidiary undertaking* is a subsidiary of a subsidiary undertaking. By reason of the link, it is described as the sub-subsidiary of the original parent.
- An *associated undertaking* is one in which a parent undertaking holds a participating interest and exercises a significant influence, without being a subsidiary undertaking or a joint venture. A typical example is Al Futaim-Wimpey (Pte) Ltd incorporated in the United Arab Emirates, which is an associated undertaking of George Wimpey plc.

A group consists of a parent company and all its subsidiary, sub-subsidiary and associated undertakings, as illustrated in Figure 1.1.

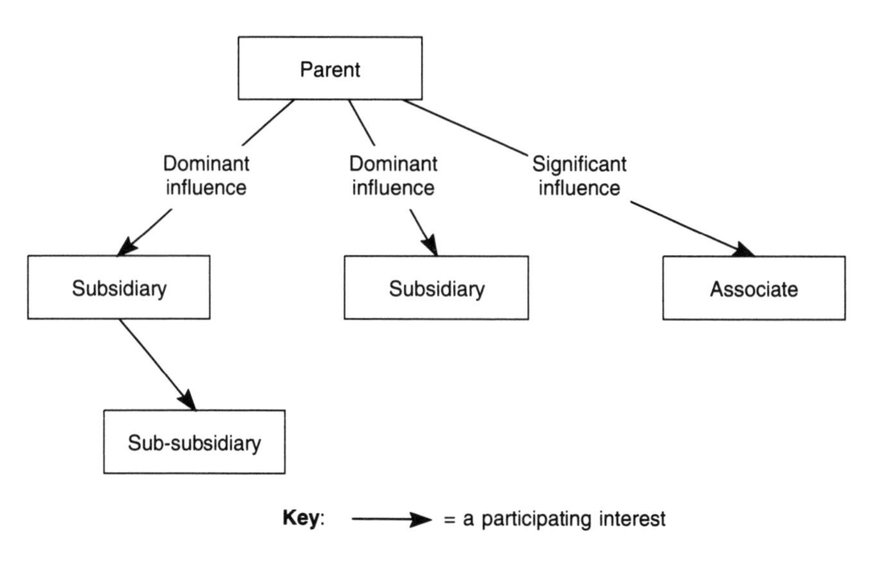

Figure 1.1 Group structure

- *Minority interests* is the label applied to the value of the interest in the net assets of the group, attributable to the shares held by non-group (outside) shareholders.

1.7 Financial reporting

All businesses have to keep financial records. In the case of limited companies the form and content is largely prescribed by law, as will be seen in Chapters 3 to 6. Apart from this constraint, it would not be possible for a business to operate properly or at all if no records were kept. Even in the case of sole proprietorships and partnerships (see Section 1.2) where record-keeping is not a legal requirement, it is a practical necessity for the reason given earlier, including the fact that the Inland Revenue, via HM Inspectors of Taxes, requires the production of properly prepared accounts in order to make an assessment for tax; similarly, HM Customs and Excise requires financial returns for Value Added Tax (VAT) purposes. Profits of unincorporated businesses are chargeable to income tax; those of limited companies are liable to corporation tax.

2 Sources of Business Finance

2.1 Introduction

There are various ways in which businesses finance their existence and operation. Not all means are available to all businesses. This chapter looks at those means of financing which are exclusive to particular forms of business formation described in Chapter 1 and also at those available to all businesses.

2.2 Sole proprietorship

At the outset, the owner contributes an amount of cash from his personal savings by transferring it to a bank account opened in the name of the business. The amount involved is termed his *capital*. Capital in this context is a specialist use of the term. It must not be confused with other meanings of this word – main, principal or excellent (a 'capital' idea), for example, or with the meaning of capital in Economics. In the accounting context, capital represents the owner's stake in the business and is the source from which items needed in the running of the business can be bought. Thus, at the outset of a new business, capital is represented solely by the cash and/or bank, balance but this is then used to acquire essential items such as premises, a vehicle, fixtures and fittings and a stock of items to resell. At this point, capital is represented by these items which have been exchanged for cash, together with the cash and bank balance remaining.

2.3 Partnership

In this case, some or all of the individual partners contribute varying sums to the partnership's cash and bank accounts as capital. After this initial cash injection, capital is subsequently held in different forms, as in Section 2.2 for sole proprietors.

Some of the partners may also lend money to the *firm*, as a partnership business is called, for a limited or indefinite period of time. The difference between a loan and a capital contribution is that, in the event of a partnership being terminated, partners' loans are repaid in priority to capital and, during the period of the loan, may bear interest. In similar fashion to capital, the money loaned is exchanged for items needed for running the business.

2.4 Limited companies

It was stated in Section 1.3 of Chapter 1 that the members or owners of a limited company are termed shareholders. The capital of a limited company is contributed, or subscribed, by the shareholders and is known as *share capital.* The company may also borrow money by means of *loan capital.* Details of the most common forms of share and loan capital now follow.

2.5 Types of share capital

The two most common types of share capital are ordinary shares and preference shares.

- *Preference shares* are so called because they are preferential in various respects. They carry the entitlement to a *fixed annual dividend,* usually expressed as a percentage of their nominal value. They are frequently designated as *cumulative*: that is, in the event of there being insufficient legally available profits in a particular year, unpaid preference dividends are carried forward until such time as they can be paid, as arrears, along with that current year's dividend. Many preference shares take priority of repayment on liquidation of the company. Some preference shares may also be designated as *convertible,* meaning that they are capable of being converted into ordinary shares on the terms specified when they were first issued. Slough Estates plc, the property investment and development company, has what are described as cumulative redeemable convertible preference shares in issue, all of which will have been redeemed or converted by 1 September 2011. The term *redeemable* means that when these shares were first issued, the company claimed the right to redeem them, that is, to buy them back from their holders (as opposed to buying them back via the Stock Exchange) at the company's option at various times and prices. Preference shares do not normally give their holders the right to attend and vote at general meetings. The nominal (or face) value of preference shares is usually either 100, 50 or 25 pence per share, but could be any other such amount as the company decides.
- *Ordinary shares* are the most common of all share issues and are frequently the only type of share which a company issues. They do not carry any rights to a fixed dividend. Dividends on ordinary shares, also known as *equity shares* or simply as *equities,* often vary according to the amount of legally distributable profit available after payment of preference dividends. In the event of a company being wound up and ceasing to exist, equity shareholders are the last to be paid and thus have to receive a share of whatever remains. By reason of this factor and the

volatility of the dividends, ordinary shares are regarded as the risk capital of a company. Although some ordinary shares can be issued as non-voting shares, they normally carry voting rights at the rate of one vote per share. The nominal value of ordinary shares is typically 100, 50, 25 and 5 pence per share but could be any such amount as the company decides. The accounts of John Maunders Group plc, the residential property development undertaking, which appear in Appendix I, show that the share capital consists entirely of ordinary shares of 20 pence per share. Slough Estates plc, cited earlier in connection with preference shares, also has ordinary shares of 25 pence per share in issue. Redland plc, manufacturer of roof tiles and aggregates for the construction industry, and George Wimpey plc, both have a share capital consisting entirely of ordinary shares of 25 pence per share.

2.6 Types of loan capital

There are several types of *loan capital*, that is, of public borrowings for a period in excess of one year, some of which are very sophisticated and complex, but the two types commonly encountered are *debentures* and *convertible loan stock*.

- *Debentures* are a form of borrowing issued in fixed units, usually of £100, which carry a fixed rate of interest and are almost invariably issued at a discount. Thus £1 000 000 8 per cent redeemable debentures 2007–2013 at 96 means that a nominal one million pounds worth of debentures has been issued at £96 per £100 nominal, thus resulting in a cash borrowing of £960 000. Interest payable, irrespective of whether profits are available or not, is 8 per cent of £1 000 000 nominal, that is, £80 000 per annum. The debentures can be redeemed by the company direct from the debenture holders, gradually or in a single operation within the years 2007 to 2013, either at par (that is, at 100) or at a premium, say 103, depending on the terms of issue. The reason why a company issues redeemable debentures (or shares) is that it may need substantial finance for only limited periods of time, such as when it starts up or when it expands. After a while, the extra finance is no longer needed and so the company redeems and cancels it. Debentures may be issued as debenture stock, that is, as fractional amounts. Most debentures are secured, giving the holders priority repayment rights over other parties to whom the issuing company owes money, by means of fixed or floating charges enforceable by debenture holders' trustees in the event of the company defaulting on payment of interest or principal at the stipulated times.

A *fixed charge* is secured on specific possessions of the company, frequently premises, and is very similar to a mortgage on property. When charged, the company's premises cannot be disposed of without the debentures first being redeemed, or bought back on the open market, and cancelled. If the company should default in paying interest when due or in repaying principal, the trustees can seize the premises charged, sell them, pay the debenture holders whatever is owing to them, deduct the costs of enforcing the charge and then hand over any remaining proceeds to the company.

A *floating charge* is non-specific in that it covers the company's possessions generally and only crystallises if the company defaults. Until that occurs, there is no restriction on the trading of these possessions. If default in paying interest or repaying principal occurs, the trustees can appoint a receiver whose powers supersede those of the directors. The receiver then sells such of the company's possessions as are necessary to discharge the amounts owing to the debenture holders and his own costs and fees. Slough Estates plc has in issue 11¼% first mortgage debentures (secured), redeemable in 2019.

- *Convertible loan stock* is another form of loan capital but is unsecured. It is similar to debentures in that it carries a fixed rate of interest, is redeemable and is quoted in units of £100. The extra feature, however, is that it can be converted into ordinary shares of the issuing company at a varying rate of number of ordinary shares per £100 of loan stock according to the date conversion takes place.

2.7 Other forms of finance

Apart from the financing methods so far considered, there are a variety of other forms available to all businesses, whether incorporated or not, the most common of which are now explained.

- *Loans* for short-term (1 to 3 years), medium-term (3 to 7 years), and long-term (over 7 years) can be obtained from individuals or institutions, principally banks and loan corporations. The loans may be secured or unsecured and carry a fixed rate of interest.
- *Bank overdraft* is a facility granted to a customer to draw out of their account a greater amount than the amount they are in credit. There is a ceiling on the amount which can be overdrawn, and this may be secured or unsecured. Interest is charged at rates which vary with bank base lending rates. Technically the bank can call the overdraft in at very short notice. This may happen but many small businesses still regard overdrafts as semi-permanent forms of finance.

- *Mortgages* are specialised forms of loan secured on property at fixed or variable rates of interest. If the mortgagor (borrower) defaults on repayments of interest or on principal repayments, the mortgagee (lender) can foreclose. This means that the property can be sold and the mortgagee repaid from the proceeds; any surplus is given to the borrower.
- *Purchasing on credit* is the normal procedure for any business, whereby goods and services are obtained on normal trade terms, under which suppliers provide the goods or services without immediate payment.

  The items are subsequently invoiced to the business consumer and settlement takes place at a pre-agreed time, typically within 14, 21 or 28 days, or by the end of the following month.
- *Hire purchase and credit sale agreements* are similar, but not identical, methods. Both involve the supply of goods for which payment is made in a series of instalments, on a monthly or quarterly basis, over a period of time. The hire purchase and credit sale prices are higher than the price which would have applied in the case of a normal cash or credit transaction, to include an interest element. The point of difference between the two types of agreement is that under a credit sale agreement the legal ownership of the goods passes from the supplier to the buyer at the outset of the contract, but only after the final instalment has been paid in the case of a hire purchase contract.
- *Finance leases* are similar to hire purchase agreements in that goods are supplied in exchange for regular rental payments over a period of years. However, the ownership of the goods always remains with the supplier (or its finance company) and never passes to the business which has possession and use of the items, except as a separate arrangement. There are taxation advantages for both parties with this arrangement and it is a very common method of financing business fleet vehicles.

3 Financial Views of a Business

3.1 Introduction

The financial reports published by businesses are of interest to a variety of individuals and organisations, for various reasons. This chapter identifies the main interested parties and reports, and the regulatory framework within which they are prepared.

3.2 Main users of financial reports

Financial reports, identified in Section 3.3 and explored in detail in Chapters 4, 5 and 6, are scrutinised and evaluated from widely differing perspectives by numerous entities and individuals who have actual and potential connections with the business, as now shown.

Lenders are particularly keen to know whether the business will be able to meet the interest payments during the period of the loan and to repay the principal sum at the end. In technical terms, lenders examine short-term liquidity – the ability of a business to meet its day-to-day commitments – and long-term solvency – the adequacy of cash funds in the years ahead. These aspects are covered in Chapter 8.

Suppliers of goods and services are also interested in short-term liquidity because they would not be prepared to despatch supplies unless the business could pay for them. In addition, if they are substantial suppliers, they will want to examine the long-term strength and prospects of the business because, if it should fail, the suppliers could be left with large quantities of stock for which alternative outlets could not be found.

Customers need to be assured that the business is sufficiently sound financially, such that the orders they have placed will be fulfilled on time. This is of particular concern to major customers who rely on a business for a large proportion of their supplies. If, because of financial failure of the business, supplies to major customers were to be disrupted, this might have catastrophic consequences. Faced with a sudden, unexpected loss of supplies, the customers would have to try to arrange alternative sources – no easy task at short notice – and probably at exorbitant prices, or have their production brought to a complete standstill through lack of essential materials or components.

Investors, their advisers and business analysts examine financial reports from several angles. Their respective interests will focus on those aspects which shed light on their investment policies. Many investors require a reasonably regular, and possibly increasing, return on their outlay as a contribution to their total income. This is particularly important where, as in the case of retired people, they are not receiving a regular wage or salary. At the same time, however, they want to see whether the underlying value of their investment will not merely be secure and stable in value but will exhibit capital growth over a period of time. Somewhat different considerations apply where an investor is considering acquiring a financial stake in a business whereby control of that business can be gained. Under these circumstances the investor is able to control the return on the investment and to exert a large degree of influence over the capital growth. Both classes of investor are interested in the financial records of the business to date, but more so in future prospects. The latter is largely subjective – a matter of opinion based on informed guesses, forecasts and estimates.

Government departments are concerned with the financial reports of business, for a variety of reasons. The involvement of the Registrar of Companies was noted earlier, in Chapter 1, Section 1.3. The Inland Revenue assesses and levies income tax on the profits of unincorporated businesses, and corporation tax on those of companies, using financial reports as a basis in both cases. Value Added Tax (VAT) is collected by the HM Customs and Excise on the basis of returns submitted by all registered businesses. Prices and production figures are collected and published regularly by the Central Statistical Office, together with their derivatives – indices, for example – and a census of production is compiled.

Employees take an interest in the financial reports of their employers, for two main reasons. First, job security is bound up with the financial strength and stability of the employer. Second, the profitability and general financial health of the business are factors which can be taken into account when formulating a pay claim.

Competitors are always eager to gain as much information as possible about competing businesses in order to secure their own survival and to promote their own growth. These interests are equally valid if they are contemplating one of the forms of business combination detailed in Chapter 1, Section 1.5.

3.3 Financial view of the business

The user groups identified in the previous section are interested in one or more financial views of a business. For interested parties, the only accessible financial information is that contained in the Annual Report and Accounts, a document which all companies are compelled to produce, publish and deposit with the Registrar of Companies, where it can be inspected by anyone. Extracts from the Annual Report and Accounts of the John Maunders Group plc are reproduced in Appendix I. Apart from financial data which may be included as an optional extra – 5- or 10-year summaries and other items – there are three financial reports which are obligatory, each dealing with a different aspect of the business. These are the *balance sheet*, the *profit and loss account* and the *cash flow statement*.

- *The balance sheet* shows the financial position of a business at a particular point in time. In simplistic terms it can be likened to a still photograph of the value of what the business owns and what it owes. Some or all of the values attributed to what the business owns may be objective (that is, based on what they actually cost when acquired in the past) or subjective (arrived at by informed guesswork) to show their current worth. The balance sheet of the John Maunders Group plc (Appendix I) reveals that some items, including computer equipment, plant and equipment, and vehicles are shown at cost while investment properties are revalued periodically to a current open market value. The form and content of the balance sheet is the subject matter of Chapter 4.
- *The profit and loss account* discloses the profit or loss made by the business over a specified length of time. Individual items that have contributed to the overall profit or loss are marshalled and arrayed in such a way as to highlight intermediate stages prior to the final figure of profit or loss. Appendix I contains the profit and loss account for the John Maunders Group plc for the year ended 30 June 1993. Component items of profit and loss accounts and their format are covered in Chapter 5.
- *The cash flow statement* is a classified schedule showing the sources of the cash funds of a business and the manner in which they have been utilised. The content and format of cash flow statements is dealt with in detail in Chapter 6. The cash flow statement of the John Maunders Group plc is shown in Appendix I. Like the profit and loss account, the cash flow statement covers a period of time.

3.4 Constraints in preparing financial statements

The three main published financial statements previously noted are prepared under a regime of regulatory requirements laid down by UK law, Accounting Standards Board pronouncements, European Union (EU) directives and UK Stock Exchange rules, each of which are now further explained.

UK Law The main constraints surrounding the preparation and presentation of the published accounts of companies are contained in the Companies Act 1985, as modified by a further Act in 1989 and by several Statutory Instruments made by the Secretary of State. The Companies Act requires all companies to produce and publish an Annual Report and Accounts to include a Directors' Report, Profit and Loss Account, and Balance Sheet. The formats of these accounting statements are specified, together with the detailed content of their supporting notes. These will be dealt with in Chapters 5 and 6.

Accounting Standards Board (ASB) This was established in 1990 to replace the existing Accounting Standards Committee (ASC). The aims of the ASB include the establishment and improvement of standards of financial accounting and reporting. One method by which it achieves this objective is the issue of Financial Reporting Standards (FRS) which specify the accounting treatment to be accorded to a given transaction or situation and thereby outlaw alternative treatments or approaches which would otherwise be available. In other words, FRSs aim to standardise accounting treatments wherever possible. The ASB at vesting date took over the corresponding pronouncements, known as Statements of Standard Accounting Practice (SSAP), from the ASC and adopted them so that they have the same validity as FRSs. The ASB has its own enforcement agency, the *Financial Reporting Review Panel (FRRP)*, which monitors companies' compliance with SSAPs and FRSs. In the event of failure of a company to comply with a standard practice, the FRRP can order the company to do so. If the company refuses to carry out the instruction to withdraw its published financial statements and to reissue them in an amended form, the FRRP can seek an Order of the High Court, and compel the company to do so. The directors of the company would then be liable personally for meeting all the legal costs of the FRRP as well as those of their own company.

An offshoot of the ASB is the *Urgent Issues Task Force (UITF)* which deals with situations where there are unsatisfactory or conflicting interpretations of existing Companies Act provisions or accounting standards (SSAPs and FRSs). The consensus decisions of the UITF are issued as rulings in the form of Abstracts and have the same force as an accounting standard.

European Union (EU) Directives These have to be embodied in the legislation of member states. The EU's Fourth Directive stipulating the format of company profit and loss accounts and balance sheets was assimilated into the Companies Act 1981 passed by the UK Parliament (but since consolidated into the Companies Act 1985); the Seventh Directive, which was concerned with certain matters pertaining to the published accounts of groups of companies, was subsumed under the Companies Act 1989 in the UK.

UK Stock Exchange Rules These apply to those companies whose shares are listed (quoted) on the Stock Exchange. The rules for annual published accounts are similar to those contained in the Companies Acts but in some particulars are more prescriptive. Another area covered by the rules is the form and content of reporting accountants' reports, which have to be included in a Prospectus when a company proposes to make a share issue.

4 Financial Position and the Balance Sheet

4.1 Introduction

All businesses need resources in order to operate; these resources are provided partly by the owners and partly by outsiders. As previously stated, the balance sheet shows the resource position of a business in financial terms at a specified moment of time, in the manner of a still photograph. This chapter examines the terminology, form, content and basis of valuation of balance sheet items commonly encountered by reference, where appropriate, to the balance sheet and notes of the John Maunders Group plc reproduced in Appendix I.

4.2 Assets

The resources, mentioned in Section 4.1, which a business owns and uses in the process of generating its income, are termed *assets*. There are two basic categories of asset: *fixed* and *current*.

- *Fixed assets* are those which require a substantial outlay but which last for more than one financial year. Effectively, they constitute much of the productive capacity of the business, thus the majority of fixed assets are tangible but others are investments and/or are intangible.
- *Tangible fixed assets* are the physical resources employed by the business. Land, buildings, plant, equipment and vehicles are the tangible fixed assets of the Maunders Group as shown in Note 11 on p. 195. They are included in the balance sheet on the basis of their historical cost: that is, at their original cost minus depreciation, which is the measure of their loss in value due to passage of time, wear and tear, obsolescence and other factors. The resultant figure is known as their net book value (NBV) or, alternatively, their written-down value (WDV). The investment properties are, however, valued on the basis of their open market existing use, this being considered most appropriate under the circumstances.

  An interesting point is raised by the inclusion, under the heading of plant, of items acquired under hire purchase agreements. In the opening paragraph of Section 4.2, it states that assets are those resources which a business not only uses but also owns. In Chapter 2, Section 2.6 we have already seen that, in the case of hire purchase agreements, the ownership of items remains with the suppliers until the final instalment has been paid at the end of the contract. On this basis it would seem that the Maunders Group is wrong to include, as assets, items which it uses but does not yet own. In fact, its treatment is correct because, for accounting purposes, the commercial reality of the situation is regarded as more important than the strict legal position. The principle of the precedence of the substance of the transaction over its legal form is applied to justify the inclusion, as fixed assets, of items in process of being acquired under hire purchase agreements. This same principle extends also to items held under finance lease contracts dealt with in Chapter 2, Section 2.6.

- *Investments* (that is, shares and/or debentures held in other companies) are classed as fixed assets if they are to be held over a long term by the investing company. Chapter 1, Section 1.5 covered the situation where a controlling interest can be gained via a shareholding, and Section 1.6 was concerned specifically with the acquisition of shareholdings constituting a participating interest coupled with a dominant influence, whereby the investee companies are subsidiary undertakings of the investing parent. Note 12 to the accounts of the Maunders Group (see p. 195) indicates that the parent company held investments in eight subsidiary undertakings.
- *Intangible fixed assets* are non-physical assets with a long-term value. Examples include rights of various sorts and goodwill, valued in each case at cost of acquisition minus any subsequent amortisation. (*Amortisation* is the term used for the depreciation of intangibles). Commonly encountered intangibles are patents – the exclusive right to manufacture an invention or to use a special process, and trade marks – the exclusive right to attach a commercial name to a product as a hallmark of excellence. Goodwill is the excess of the purchase cost of another business, or a controlling interest in it, over the book value of the investment. The only intangible asset disclosed in the Maunders Group balance sheet is the capitalised value of ground rents on sale of properties, representing a valuation of seven years' rent receivable.
- *Capital expenditure* is the term applied to money expended on the acquisition of fixed assets.
- *Current assets* are those assets which a business owns and uses within the annual operating cycle. They may be physical assets which will ultimately be converted into cash, or a legal right to receive cash. Common current assets are stocks – of raw materials (for conversion into finished products), work-in-progress (or process), partly completed products, finished goods awaiting sale or resale, and debtors (the amounts owing to a business by its customers who have bought goods and/or services but have not yet paid for them). As might be expected of a property developer, the balance sheet of the Maunders Group identifies

as current assets: land held for development; stock of building materials; work-in-progress (properties in course of construction); debtors; cash in hand, and cash held by the bank. Showhouses are included under development land and stocks.

4.3 Liabilities

This term is used to denote the financial obligations of a business to pay for goods and services supplied to it on credit, to repay borrowings, and to pay over to the taxation bodies sums collected on their behalf by the business in an agency capacity, together with amounts levied on profits. All these obligations are *external liabilities*, so called because they are amounts owing to individuals and entities outside the business. There are, however, *internal liabilities*, by which is meant the obligations of the business to its owner(s) for the capital invested and which is returned to the owner(s), plus any undistributed profits which the business has made, or minus any losses which it has suffered, when the business is terminated.

External liabilities are categorised as *current*, meaning due for settlement in the short term, under the heading *'Creditors: amounts falling due within one year'*; and as *long-term*, under one of two headings: *'Creditors: amounts falling due in more than one year'*, and *'Provisions for liabilities and charges'*.

- *Creditors: amounts falling due within one year* are short-term liabilities to be found in the balance sheets of all businesses. The most common ones are trade creditors, corporation tax on profits, VAT, income tax and national insurance contributions, proposed dividends, accruals (accumulating expenses), and bank overdrafts. Many businesses regard a bank overdraft as a semi-permanent (and therefore, long-term) variable source of finance but, because it is technically repayable on demand or at short notice, it is classed as a current liability. All the above listed short-term liabilities are to be found in the balance sheet of the Maunders Group, along with amounts owed to subsidiary undertakings, instalments due under hire purchase contracts and short-term loans.
- *Creditors: amounts falling due in more than one year* does not usually include trade creditors as they would not ordinarily be willing to wait that long for settlement. Items typically found are medium-term and long-term borrowings, hire purchase instalments, and rentals payable under finance lease contracts, but excluding, in each case, the future interest/finance charge elements.
- *Provisions for liabilities and charges* include sums charged against profits for events which have occurred where the amount involved cannot, at the end of the financial period, be calculated with any degree of precision. Examples of provisions include those to meet legal costs and

damages for breach of contract or negligence, and to meet charges for corporation tax, payment of which can be postponed. This latter, provision for deferred taxation, is disclosed in Note 17 of the Maunders Group balance sheet (see p. 198), together with a provision for warranties – the contractual obligation to rectify building defects, currently latent, which become apparent with the passage of time.

Internal liabilities, to which reference has already been made comprise *share capital* and *reserves.*

- *Share capital* of companies has been described in Chapter 2, Section 2.5. As noted in Chapter 1, Section 1.3, the number and denomination of each type of share capital which the company is empowered to issue is declared in the company's Memorandum of Association and is known as the *authorised capital.* A company may not need to issue the whole of its authorised capital at any given time. The nominal value of the aggregate of those shares which have been issued is termed the *allotted share capital.* Most companies have taken powers to call up shares by instalments: in other words, to require full payment over a period of time. The total amount called up in aggregate is referred to as *called up share capital.* From this it follows that the figures for authorised, allotted and called up share capital may differ. Note 18 of the Maunders Group balance sheet (see p. 198) reveals that the company has allotted less than its authorised maximum, but that the whole of the allotted share capital has been called up and is fully paid, meaning that none of the shareholders has defaulted in paying the sums called up.
- *Reserves* are the accumulated undistributed profits of the company. The existence of reserves does not mean that they can in fact be distributed in the normal course of events. There are severe legal restrictions on certain types of reserve. The Companies Act 1985 prohibits the distribution, as dividends, of unrealised profits and of the balance of the share premium account. Unrealised profits arise, say, when asset values are raised by revaluation; the revaluation surplus is, in effect, a profit which exists only on paper, but it can be converted into a realised profit (and therefore becomes distributable) if the revalued assets are subsequently sold.

Share premium is the term used to describe that part of the price at which shares are sold (by the company) which exceeds the nominal value. This premium, or excess, must be kept in a separate account and can be utilised for a very limited number of purposes.

Note 19 of the Maunders Group balance sheet (see p. 198) shows three reserves – share premium, revaluation and profit and loss. Of these, only the profit and loss account can be regarded as distributable.

4.4 Balance sheet equation

It is a truism that all the resources which a business owns and uses (assets) can only have been acquired from finance provided by outsiders (external liabilities) and owners (share capital and reserves). This fact is expressed in what is termed the balance sheet equation which states that, at all times,

Aggregate assets equal aggregate liabilities (internal and external).

Proof of this can be obtained by an examination of the John Maunders Group plc balance sheet for the group, for 1993:

	£000
Fixed assets	2 153
Current assets	42 785
Total assets	44 938

	£000
External liabilities	
Creditors	
due within one year	18 525
due in more than one year	413
Provisions for liabilities and charges	93
Total	19 031
Internal liabilities	
Capital and reserves	
(Shareholders' funds)	25 907
Total liabilities	44 938

4.5 Other balance sheet terminology

Apart from the terminology encountered already, further items need to be identified and their meaning and significance explained, again by reference to the Maunders Group balance sheet.

- *Net current assets* is a figure which shows by how much current assets exceed current liabilities. It indicates the extent to which the short-term external obligations of the business can be met out of existing cash and bank balances, and items such as stock and debtors which are convertible into cash in the short term. It is technically possible for the figure to be negative, in which case the label is altered to *net current liabilities*, meaning that the business is unable to discharge its external obligations due for settlement in the immediate, or very near, future. Businesses which find themselves in this situation are in severe financial straits and are usually on the verge of collapse. Net current assets for the Maunders Group for 1993 are:

	£000
Current assets	42 785
less	
Current liabilities	
(Creditors due within one year)	18 525
Net current assets	24 260

- *Working capital* this means precisely the same as net current assets, by which it has been replaced in official terminology.
- *Net assets* is a figure which shows how much of the book value of all the assets of a business is left over after all external liabilities have been taken into account. Alternatively, it can be viewed as the aggregate of the fixed and net current assets from which external long-term liabilities have been deducted. Yet another aspect of this figure is that it indicates the extent to which the book value of the total assets has been financed by the shareholders. This is explained further in shareholders' funds that follow. The net assets figure for the Maunders Group for 1993 is arrived at as follows:

	£000
Fixed assets	2 153
Current assets	42 785
	44 938
less	
External liabilities	
Creditors	
due in less than one year	18 525
due in more than one year	413
Provisions for liabilities and charges	93
	19 031
Net assets	25 907

Using the alternative view, it represents:

	£000
Fixed assets	2 153
Net current assets	24 260
	26 413
less	
Long-term external liabilities	
Creditors due in more than one year	413
Provisions for liabilities and charges	93
	506
Net assets	25 907

A popular misconception is that the net assets figure shows what the business is worth, but this is not so. Value in an elusive concept. In any transaction the buying and selling parties each have their own ideas of what the object concerned is worth. The eventual price is settled after negotiation and compromise. A business, however, is ordinarily regarded as a going concern – that is, as an indefinitely continuing entity. As a consequence, any attempt to introduce sale values is irrelevant in this context.

Businesses deal with this situation by basing their asset values on objective, factual figures or, in other words, what the assets cost in the first place, but reduced, where necessary, by losses in value related to such factors as wear and tear, obsolescence and, in the case of trading stocks, where the resale price has fallen to below cost.

It is not uncommon, however, for companies to revalue their fixed assets when it seems appropriate to do so. This situation can arise when a long-established company, which acquired its premises many years before, wishes to reflect current values in its balance sheet. The valuation is carried out professionally and the fixed assets appear in the balance sheet at the revalued figures, with the uplift appearing in reserves as already described in this Chapter (in Section 4.3). This practice is, however, fraught with danger. The classic example concerns Queens Moat Houses plc, the hotel group. In December 1991 property valuers Weatherall Green and Smith valued the group's hotels at about £2 billion. It was then announced in October 1993 by Queens Moat Houses that a new valuation in December 1992 by a different firm of valuers, Jones Lang Wootton had produced a figure of £861 million, a fall of well over £1 billion in one year. This had a catastrophic effect on the company's financial position. The property devaluation, in conjunction with other losses, resulted in negative net assets (that is, net liabilities) of £380 million, and a near-record high loss of any UK company, of over £1 billion; it rendered the company's shares, which in early 1993 were being traded at 59 pence per share, virtually worthless. At their peak in late 1989, they had traded at almost 120 pence per share.

This affair culminated in the Royal Institution of Chartered Surveyors (RICS) issuing revised Statements of Valuation and Appraisal Practice and Guidance Notes in 1994, and including relevant recommendations of the Mallinson Committee.

- *Shareholders' funds* is the term applied to the aggregate of called-up share capital and all distributable and non-distributable reserves, the entire amount of which would be shared out pro rata between the ordinary shareholders in the event of the company being wound up. As has already been noted, it is the shareholders' funds which finance the net assets and for that reason the two figures are identical (at £25 907 000 in the case of the Maunders Group in 1993).

4.6 Balance sheet format

The balance sheet of the Maunders Group plc reproduced in Appendix I is a good example of a reasonably straightforward structure. More complex examples can be found in practice, as illustrated by the example shown in Table 4.1.

Table 4.1 Balance sheet format

X Y Z Group plc

Consolidated balance sheet
as at 31 December Year 5

	£000	£000	£000
Fixed assets			
Intangible assets			
Development costs	X		
Patents, trade marks	X		
Goodwill	X		
		X	
Tangible assets			
Land and buildings	X		
Plant and machinery	X		
Fixtures, fittings, tools and equipment	X		
		X	

Investments			
Interests in associated undertakings	X		
Other participating interests	X		
Other investments other than loans	X		
Other loans	X		
		X	
			X
Current assets			
Stocks			
Raw materials and consumables	X		
Work in progress	X		
Finished goods and goods for resale	X		
		X	
Debtors			
Trade debtors	X		
Amounts owed by undertakings in which the company has a participating interest	X		
Other debtors	X		
Prepayments	X		
		X	
Investments		X	
Cash at bank and in hand		X	
		X	
Creditors: amounts falling due within one year			
Debenture loans	X		
Bank loans and overdrafts	X		
Trade creditors	X		
Amounts owed to undertakings in which the company has a participating interest	X		
Other creditors including taxation	X		
Proposed dividends	X		
		(X)	
Net current assets			X
Total assets less current liabilities			X
Creditors: amounts falling due after more than one year			
Debenture loans		X	
Bank loans		X	
Trade creditors		X	
Other creditors including taxation		X	
			(X)

Provisions for liabilities and charges		
Deferred taxation	X	
Other provisions	X	
		(X)
Minority interests		(X)
Net assets		X
Capital and reserves		
Called up share capital		X
Share premium account	X	
Revaluation reserve	X	
Other reserves		
Capital redemption reserve	X	
Profit and loss account	X	
		X
Shareholders' funds		X

5 Profit Measurement and the Profit and Loss Account

5.1 Introduction

In addition to establishing the financial position of a business at particular points in time, the owners need to know whether the business is operating at a profit or a loss, and by what amount. For legal reasons, all limited companies have to produce and publish the Annual Report and Accounts for each financial year, which includes the profit and loss account and balance sheet. The profit and loss account spans the whole financial year and links the balance sheet at the end of the previous year with that at the end of the current year. This chapter examines the terminology, form, content and theoretical basis of items commonly encountered by reference, where appropriate to the profit and loss account and notes of the John Maunders Group plc reproduced in Appendix I.

5.2 Profit and loss

The determination of profit and loss is carried out by comparing revenue and expense. If the revenue of a period exceeds the expense, the result is a profit; if expense is greater than revenue, a loss has occurred. There are various points at which profit or loss is measured, which are considered in Section 5.6 of this chapter. Before that point is reached it is necessary to examine the nature and basis of revenue and expense.

5.3 Revenue

In the context of the profit and loss account, the term *revenue* simply means the sales value of goods manufactured and/or sold. In businesses which neither manufacture nor sell merchandise, revenue consists of the value of the services they have provided to their clients. Thus the revenue of a manufacturer is comprised of the value of its production invoiced to its wholesalers, whose revenue is the amount they invoice to retailers. It follows that retailers' revenue is what they obtain by selling goods to the public. The revenue of a building/property development company consists of the aggregate of the selling prices of domestic and industrial premises sold together with such amounts as reflect the value of work carried out on long-term contracts. The fees of a firm of architects or civil engineering consultants constitute their revenues. The revenue of a business is split into one of two categories – *turnover* and *other revenue*.

- *Turnover* is the descriptive label applied to the main source of revenue of a business. The nature of the business determines whether an item of revenue is classed as turnover or not. In the circumstances of the manufacturer, wholesaler and retailer described in the previous paragraph, the whole of the revenue referred to would be classed as turnover. If, however, any of these three types of business let any surplus floor space to an outside occupier on a landlord/tenant basis, any rent received would be treated as 'other revenue'. On the other hand, rents received by a property investment company, Slough Estates plc, for example, would constitute turnover. Similarly, rentals received from the hiring out of plant and equipment by a plant hire company would be treated as turnover, hiring of plant being the main objective for which the company was established in the first instance.
- *Other revenue* describes all other items of revenue deriving from sources other than the main activity of the business. Examples of items of other revenue in a manufacturing and/or merchandising business include rent received (already noted above); discounts received for prompt payment of outstanding invoices; dividends and interest received from investments; hire purchase interest received; and leasing rentals received.

A question arises as to the point at which revenue is deemed to come into being. This is a very important point because, as has been stated in Section 5.2, profit or loss is derived by comparing figures for revenue and expense. There are special rules for certain circumstances but the general rule is that revenue is recognised, that is, comes into existence, when the sale transaction takes place. Where goods are sold for cash, as in the retail trade, the sale occurs and the revenue is recognised when the cash is handed over in exchange for the goods. Many transactions, however, are carried out on credit terms. In these instances, the sale is deemed to occur when the invoice is raised subsequent to the handing over of the goods, or the provision of the service, and not when the money is received, which may be some weeks or months later.

5.4 Expense

This is a term applied to a cost, the benefit of which has been fully used up during the accounting year. In Chapter 4, Section 4.2 it was explained that capital expenditure is the terminology used to describe monetary outlay on fixed assets which by definition are resources lasting more than one year. This expenditure does not in itself constitute expense because its value lasts for a number of years. However, all assets have a finite life and, in recognition of this fact, a proportion of the initial cost, deemed to represent the diminution in value due to wear and tear, obsolescence and/or other causes, is treated as an expense, and labelled 'depreciation' in the profit and loss account. Expenses fall under two groupings – *cost of sales* and *other expenses.*

- *Cost of sales* is a figure which represents the cost equivalent of the turnover figure. Thus in a trading business it shows the cost of the bought-in goods which have been resold at the value indicated by the turnover figure. For a manufacturing concern, cost of sales is the manufactured cost of the items included under turnover.

  It is usual for the cost of sales figure to be derived by taking the cost of opening stock of goods and adding the cost of goods bought for resale (or manufactured for sale); from this aggregate is deducted the cost of closing stock. In arriving at the cost of bought-in goods for resale and the raw material element of goods manufactured for sale, all costs are included which have been incurred up to the point at which the items concerned have become saleable. In addition, therefore, to the basic cost (suppliers' list price) of the bought-in goods, the following costs are taken into account, as appropriate – import duties; transport charges from suppliers' premises; handling charges; and costs of converting raw materials into finished products. If applicable, trade discounts and subsidies are deducted.
- *Other expenses,* as the term implies, includes all expenses which have not been accounted for under 'cost of sales'. Examples include wages; salaries; heating, lighting and power charges; repairs and maintenance; rent; business rates; depreciation; postage; telephones; and stationery.

5.5 Accruals concept

Also known as the *accruals convention,* the *accruals postulate* and the *accruals principle,* it is one of several broad basic assumptions on which the accounts of companies are prepared.

In the context of the profit and loss account its application can be seen in the measurement of profit (or loss), which, as was stated in Section 5.2, is the difference between revenue and expense. If the accruals principle did not apply, profit or loss would be the difference between receipts from sales debtors (customers) together with cash sales receipts (if any), less payments to suppliers of goods and services.

As it stands, the operation of the accruals principle means that revenue is recognised (by inclusion in the profit and loss account) as explained in

Section 5.3, in the period in which the sale takes place, even though under normal terms of trade the amount due is received in a subsequent accounting period. In other words, the criterion for the inclusion of a sale in the profit and loss account under the heading of 'turnover' is the fact that it has been transacted in the period; the fact that the money for the sale is not due to be received until a later period is irrelevant in the calculation of turnover.

Similarly, expenses are recognised (by inclusion in the profit and loss account) in the period in which they are incurred and not in the period in which they are paid. Almost invariably, when a business obtains goods and services from outside suppliers it does so on normal business terms whereby the goods are supplied or the services are performed and the cost is invoiced by the suppliers for settlement at a later date, typically 21 days, 28 days or 2 months after the date of the invoice. The criterion for inclusion of an expense in the profit and loss account is the fact that it has been incurred in the period. The principle involved is thus consistent with that applicable in determining turnover.

A second aspect of the accruals principle is that of matching. This simply means that, in the profit and loss account, revenue and expense, in each case arrived at by application of the accruals concept, are matched to produce a resultant difference representing a profit or a loss. It follows that the cash and bank balances of the business will not be increased by the amount of profit (or reduced by the amount of loss) disclosed, because of the time lags between revenue and cash receipt inflows and between expense and cash payment outflows, as well as asset and liability cash receipts and payments.

5.6 Stages in profit and loss ascertainment

Throughout the chapter up to this point, profit or loss has been stated to be the difference between total revenue and total expense, and that the profit and loss account is the vehicle by which this operation is performed.

In practice, profit or loss is ascertained at a number of intermediate points which include:

gross profit;
operating profit;
trading profit;
profit on ordinary activities before taxation;
profit on ordinary activities after taxation; and
retained profit for the year.

Each of these is now explained by reference to the Maunders Group profit and loss account shown in Appendix I:

- *Gross profit* is the amount by which turnover exceeds cost of sales. For the Maunders Group this is shown as:

	£000
Turnover	59 809
Cost of sales	(50 774)
Gross profit	9 035

  Gross profit can be expressed either in relation to turnover, when it is termed *gross margin* or *sales margin*, or to cost of sales, when it is termed *mark up*. On the basis of the above figures, average sales margin (the gross profit content of turnover) is (9 035 x 100)/59 809, that is 15.1 per cent and average mark up is (9 035 x 100)/50 774, a figure of 17.8 per cent.
- *Operating profit* is the term used in a manufacturing business to label that part of the gross profit which is left over after expenses, other than of a financing nature, have been met. The main items of expense are grouped under the headings of *distribution costs* (if applicable, the costs of operating warehousing and transport fleet activities) and *administrative expenses* (office running costs, including wages, salaries, repairs, stationery, telephones, heating, lighting, rent and rates, together with audit fees and expenses).

  Other items accounted for in arriving at a figure of operating profit are the non-mainstream, but nevertheless important, items of rent (payable and/or receivable) and royalties (payable and/or receivable) for the exercise of mineral or patent rights or of copyright.

  The Maunders Group profit and loss account shows:

	£000
Gross profit	9 035
Operating expenses (net)	(3 740)
Operating profit	5 295

  An explanatory note discloses that the operating expenses consist of administrative expenses less operating income.

  The operating profit is thus a figure for which the sales, production and administration executives can be held responsible and be congratulated on or blamed for, according to circumstances.
- *Trading profit* (or profit from trading activities) is the term used in a merchandising business to describe the figure of profit comparable with operating profit in a manufacturing concern. Thus it is the result of deducting distribution costs and administrative expenses from gross profit, and adjusting for any items of operating income and expense. It is

used as a measure of the performance of sales, distribution and administration executives.

- *Profit on ordinary activities before taxation* is the figure of operating profit (or trading profit) after accounting for items of a financing nature and is the basis from which the taxable profit of the business is calculated. Finance charges payable and receivable include bank interest, loan interest, gross earnings and finance charges under finance lease and hire purchase, contracts together with gains and losses caused by movements in exchange rates on transactions conducted in foreign currencies. Dividends received on shares held as investments are also included, as are ground rents and rents from property lettings. In the case of the Maunders Group the figures are:

	£000
Operating profit	5 295
Investment income	78
Interest payable and similar charges	(1 536)
Profit on ordinary activities before taxation	3 837

- *Profit on ordinary activities after taxation* is the amount of profit after corporation tax liabilities have been provided. The charge for corporation tax is calculated on a figure of taxable profit which is based on, but not identical with, profit on ordinary activities before taxation. Part of the difference between a figure of corporation tax based on this latter figure and on taxable profit is as a result of phasing differences. Some companies, including the Maunders Group, counteract this by raising (or releasing) a figure of deferred taxation. The effect of this is that the charge for taxation, apart from certain exceptions where rare items of income are tax free and some items of expense are non-tax-deductible, relates directly to the figure of profit on ordinary activities before taxation. Figures for the Maunders Group are:

	£000
Profit on ordinary activities before taxation	3 837
Taxation	(1 284)
Profit on ordinary activities after taxation	2 553

- *Retained profit for the year* discloses the amount of realised profit for the year which is left over after all expenses have been accounted for, the taxation liability provided and after dividend distributions have been appropriated. This residue is sometimes referred to as *profit ploughed back*. The Maunders Group profit and loss account shows:

	£000
Profit from ordinary activities after taxation	2 553
Minority interest	1
Profit for the financial year	2 554
Dividends	(1 272)
Retained profit for the year	1 282

This is the point at which many companies terminate their profit and loss accounts. In fact, the Maunders Group has added an extra section. Explanation of this is given in Section 5.7 below.

5.7 Other profit and loss account terminology

Apart from the terminology encountered already, further items need to be identified and their meaning and significance explained, where appropriate, by reference to the Maunders Group profit and loss account.

- *Exceptional items* are items of a routine nature but of an abnormally high amount. Thus if serious structural faults were to be discovered in the office building and these necessitated costly repairs and rectification, the expense would be charged in the normal way under the heading 'administrative expenses'. Repair charges are a normal occurrence, but in the instance cited an abnormally high charge is involved. This would be regarded as exceptional and would have to be identified separately in a note to the profit and loss account. No such items have been noted for the Maunders Group.
- *Extraordinary items* are items which are abnormal by being both very significant in amount and very rare and non-recurring. For this reason, actual examples are difficult to find. A hypothetical example is the circumstance where an employee is fatally injured on factory premises and the deceased's widow sues the company successfully for damages. If the company is inadequately insured against such an occurrence, the very substantial amount of damages and legal costs (of both parties) would constitute an extraordinary item of expense.

  In the profit and loss account, extraordinary items are located between the headings 'profit on ordinary items after taxation' and 'profit for the financial year' and are disclosed net of taxation effects. No extraordinary items have been noted in the Maunders Group profit and loss account.
- *Prior year items* are adjustments in one financial year of estimated figures used in an earlier financial year. When a company's final accounts are

being prepared, estimated amounts have to be provided when actual amounts are not available. This then results in an adjustment for the difference between the estimated and actual amounts in the financial year in which the actual amounts become known. Such adjustments affect the profit and loss account of the year in which they arise, not of the year *in respect of which* they arise. Unless the adjustments are of such magnitude as to constitute an exceptional or an extraordinary item, they are not identifiable separately. As a consequence, it is not possible to detect what prior year items have been accounted for in the Maunders Group profit and loss account.

- *Prior year adjustments* result from one of two circumstances – the financial effects of either a change in accounting policy, or the correction of a fundamental error arising in an earlier financial year, where the amounts involved in each case are substantial. The issue of a Financial Reporting Standard (FRS) could give rise to a prior year adjustment if a company has to change its policy as a result. In the Maunders Group accounting policies notes, changes have been declared in accounting policy for turnover recognition and for financing of showhouses, and the recognition of profits and losses on them. Prior year adjustments are shown in the financial year in which they arise but are given retrospective effect by amending the figure of retained profit brought forward and by amending the individual items affected in the profit and loss account and balance sheet.

  The Maunders Group has elected to effect this adjustment by extending the profit and loss account as follows:

	1993		*1992 Restated*	
	£000	*£000*	*£000*	*£000*
Retained profit for the year		1 282		1 613
Retained profit brought forward				
as previously reported	17 887		16 423	
prior year adjustment	(1 023)		(1 172)	
as restated		16 864		15 251
Retained profit carried forward		18 146		16 864

It can be seen that the new figure of retained profit carried forward in 1992 has become the restated figure of retained profit brought forward to 1993 (£16 864).

- *Profit for the financial year* is that part of the profit on ordinary activities after taxation, adjusted by extraordinary items and by minority interests. When a company holds investments in other companies such that a parent/subsidiary relationship exists (see Chapter 1, Section 1.6), the group profit and loss account discloses the amount of profit (or loss) in subsidiary companies which belongs to minority (outside) shareholders. The profit of the parent company together with that attributable to its subsidiary companies (and therefore available to the parent company's shareholders) is labelled 'profit for the financial year'. The Maunders Group profit and loss account discloses:

	£000
Profit on ordinary activities after taxation	2 553
Minority interest	1
Profit for the financial year	2 554

- *Earnings per share* is a statistic which shows how much of the profit for the financial year is available to each ordinary share. If a company has preference shares in issue, the dividend payable on them is deducted from the profit for the year figure to leave an amount of earnings attributable to ordinary shares. Earnings per share (EPS) is then expressed in pence (never as a decimal of £1). Complications can arise through various circumstances, one being when new shares are issued for cash during the year. In that event, the number of ordinary shares used as the divisor is the weighted average number of ordinary shares in issue and ranking for dividend in the year.

  This circumstance has applied in the case of the Maunders Group, which discloses an EPS figure of 10.36p. There were no preference shares in issue, but there had been an issue (for cash) of ordinary shares during the year. Consequently, the earnings figure was divided by the weighted average number of ordinary shares in issue during the year.

5.8 Profit and loss account format

The profit and loss account of the Maunders Group plc reproduced in Appendix I is a good example of a reasonably straightforward structure. More complex examples can be found in practice, as illustrated by the example shown in Table 5.1. Companies are free to adopt an alternative layout, if they consider it appropriate to do so, as illustrated in Table 5.2. From 'operating profit' onwards the format is identical with the previous one, down to 'retained profit/(loss)' for the year. This alternative format is particularly suitable for companies in the building/construction sector and is used, for example, by George Wimpey plc.

Table 5.1 Profit and loss account format

X Y Z Group plc

Consolidated profit and loss account for the year ended 31 December Year 5

	£000	*£000*
Turnover		X
Cost of sales		(X)
Gross profit		X
Distribution costs	(X)	
Administrative expenses	(X)	
		(X)
		X
Other operating income	X	
Other operating charges	(X)	
		X or (X)
Operating profit/trading profit		X
Income from interest in associated undertakings	X	
Income from other participating interests	X	
Income from other fixed asset investments	X	
Other interest receivable and similar income	X	
Interest payable and similar charges	(X)	
		X
Profit on ordinary activities before taxation		X
Tax on profit on ordinary activities		(X)
Profit on ordinary activities after taxation		X
Minority interests		(X)
		X
Extraordinary items		X or (X)
Profit/(loss) for the financial year		X or (X)
Dividends		(X)
Retained profit/(loss) for the year		X or (X)

Table 5.2 Alternative profit and loss account format

X Y Z Group plc

Consolidated profit and loss account for the year ended 31 December Year 5
[Alternative format]

	£000	£000
Turnover		X
Change in stocks of finished goods and work in progress [increase/(decrease)]		X or (X)
Own work capitalised		X
Other operating income		X
		X
Raw materials and consumables	(X)	
Other external charges	(X)	
		(X)
		X
Staff costs	(X)	
Depreciation	(X)	
Other operating charges	(X)	
		(X)
Operating profit		X

Both of the basic formats shown in Tables 5.1 and 5.2 have to be further expanded to give an analysis of each item from turnover to at least the operating profit line, to disclose separate figures for continuing and discontinued operations. Within the continuing operations analysis, figures relating to acquisitions during the period are identified separately. The analyses may be shown on a line-by-line (horizontal) basis but the simpler columnar (vertical) form of presentation is more often used. Where within a period no operations have been acquired or discontinued, the analysis is not required because the total figures constitute continuing operations.

A further requirement is that the profit and loss account has to be accompanied by three supplementary statements, namely:

1. Statement of total recognised gains and losses.
2. Note of historical cost profits and losses.
3. Reconciliation of movements in shareholders' funds.

For a general appreciation of the nature and significance of profit and loss accounts, these three statements can be disregarded as they deal with matters outside the scope of this book.

6 Sources and Uses of Cash and the Cash Flow Statement

6.1 Introduction

Without a supply of cash it would be virtually impossible for any business to function. Just as the moving parts of machinery would seize up in the absence of lubricants, so the operations of a business would grind to a halt if cash were not available to pay for goods and services and to meet everyday expenses.

It is hardly surprising, therefore, that businesses pay close attention to the cash supply. This is done in several ways, principally by preparing a classified record of where the cash has come from and how it has been used; and secondly by detailed advance planning of cash requirements. This latter aspect is dealt with fully in Chapter 16 (Cash Budgeting).

The other aspect, the classified report of cash inflows and outflows, is contained in a cash flow statement which, in common with the balance sheet and profit and loss account, dealt with in Chapters 4 and 5, is contained in the published Annual Report and Accounts. The cash flow statement, like the profit and loss account, spans the whole year and links the cash aspects of transactions in the profit and loss account with those which arise between the opening and closing dates of the balance sheet. This chapter deals with the terminology, form, content and basis of the items encountered in a typical cash flow statement by reference, where appropriate, to the cash flow statement of the John Maunders Group plc reproduced in Appendix I.

6.2 Cash flow

Except in the smallest and most simple businesses, the profit and loss for the period does not equal the net cash inflow or outflow. There are several reasons why this is so. Profit or loss is the difference between the revenues and expenses of a period, each of which is measured on an accruals basis. Thus sales revenues consist of amounts earned during the period, whether received or not and expenses are those incurred during the period, whether paid or not. Cash flows, on the other hand include all actual inflows and outflows, not only those connected with the profit and loss account, but also those which affect balance sheet items. Payments to acquire assets and to discharge liabilities are examples of the latter.

Transactions not involving cash flows – for example, the acquisition of an interest in a subsidiary undertaking for a non-cash consideration, shares and/or debentures – do not appear on the cash flow statement but are an appended note.

6.3 Cash and cash equivalents

In a profit and loss account the last figure to be deduced is that of the retained profit for the period. The corresponding figure in a cash flow statement is the movement, increase or decrease, in balances of cash and cash equivalents. The term *cash* means cash held in safes, cash boxes, cash tills and elsewhere on any of the premises of the business, together with balances held in current and deposit accounts at the bank. *Cash equivalents* is the term applied to investments convertible into cash without notice, less bank overdrafts repayable within three months from the date they were advanced.

6.4 Stages in cash flow statements

The movement in cash and cash equivalents is the difference between total inflows and total outflows of cash within an accounting period and these are disclosed in the cash flow statement.

In practice, the increase or decrease in cash and cash equivalents is ascertained at a number of intermediate points, consisting of:

- operating activities;
- returns on investments and servicing of finance;
- taxation;
- investing activities;
- net cash inflow/outflow before financing;
- financing.

Each of these is now illustrated with reference to the Maunders Group cash flow statement in Appendix I.

- *Operating activities* discloses the net cash flow (inflows less outflows or vice versa) corresponding to the operating profit figure in the profit and loss account.

  The Maunders Group cash flow statement shows:

	£000
Net cash inflow from operating activities	12 018

At first sight it may be disconcerting to see that the figure of operating profit, disclosed in Chapter 5, Section 5.6 was:

	£000
Operating profit	5 295

The very substantial difference (in the region of £7 million) between the two figures is attributable to the previously stated facts that profit is the net difference between revenues earned, whether received or not, and expenses incurred, whether paid or not. On the other hand, cash flow is the net difference between actual receipts and actual payments which relate not only to items which appear in the profit and loss account, but also to those which do not. Amounts received and paid relating to the balance sheet items are an example. Additionally, where provisions are raised in the profit and loss account, depreciation being the most common, there is no corresponding cash flow.

In order to help users of the Annual Report and Accounts to understand the reasons for the differences between the figures for operating profit and operating activity cash flows, a note has to be appended to the cash flow statement in the form of a schedule of items of differences. In the case of the Maunders Group, the schedule is given as a note to the accounts:

	£000
Reconciliation of operating profit to net cash inflow from operating activities	
Operating profit	5 295
Depreciation	305
Minority interest	(5)
Ground rents capitalised	(27)
Loss on sale of tangible fixed assets	4
Decrease in land for development	4 824
Increase in stock and work-in-progress	(81)
Increase in debtors	(701)
Increase in creditors and provisions	2 404
Net cash inflow from operating activities	12 018

Thus it can be seen that the difference between operating profit and net flows from operating activities is caused by non-cash items and to lags and leads in receipts and payments.

- *Returns on investments and servicing of finance* is a category which includes sums received by way of dividends and interest on investments and payments of dividends to shareholders and of interest to lenders, including interest on hire purchase contracts and finance charges on finance lease contracts. The Maunders Group cash flow statement discloses:

	£000
Returns on investments and servicing of finance	
Investment income received	81
Interest paid	(1 537)
Interest element of hire purchase payments	(28)
Dividends paid	(1 220)
Net cash outflow from returns on investments and servicing of finance	(2 704)

- *Taxation*, as its name implies, is the heading which includes payments to and from the taxation authorities relating to the taxable profits of the business. It does not include amounts for those items, such as VAT on sales, and income tax on interest payments, where the business merely acts as a collector. For the same reason, VAT on goods and services bought in, income tax levied on interest received and tax credits on dividends received are also excluded. Items cited as exclusions appear within the operating activities and returns on investments and servicing of finance headings. In the Maunders Group, there is only a single item under this heading:

	£000
Taxation	
Corporation tax paid	(1 364)

- *Investing activities* encompass cash flows relating to the acquisition and disposal of fixed assets and of current asset investments, other than those classified as cash equivalents. The cash flow statement for the Maunders Group shows:

	£000
Investing activities	
Payments to acquire fixed assets	(209)
Receipts from sales of tangible fixed assets	56
Net cash outflow from investing activities	(153)

- *Net cash inflow/outflow before financing* is not a sub-section itself, but a sub-total of the four preceding sub-sections, inserted to disclose the net cash flow effect, after tax, of the operating and investing activities of the business. In the case of the Maunders Group, the figures are:

	£000
Net cash inflow from operating activities	12 018
Net cash outflow from returns on investments and servicing of finance	(2 704)
Corporation tax paid	(1 364)
Net cash outflow from investing activities	(153)
Net cash inflow before financing	7 797

- *Financing* activities include receipts from, and payments to, external providers (shareholders and lenders) of finance. Amounts disclosed under this heading are restricted to the principal elements; dividends paid and interest paid have already been stated and appeared under the returns on investments and servicing of finance classification as also have interest paid and finance charges paid on hire purchase and finance lease contracts respectively. The figures of the Maunders Group are:

	£000
Financing	
Capital element of hire purchase payments	(237)
Decrease in secured loan	(427)
Issue of ordinary share capital	46
Net cash outflow from financing	(618)

6.5 Cash flow statement format

The cash flow statement of the Maunders Group plc reproduced in Appendix I is a good example of a reasonably straightforward structure. More complex examples can be found in practice, as illustrated by the example in Table 6.1. The statement is followed immediately by a series of notes, shown in the example given in Table 6.2, all of which are obligatory.

Table 6.1 Cash flow statement format

X Y X Group plc

Consolidated cash flow statement for the year ended 31 December Year 5

	£000	*£000*
Operating activities		
Cash received from customers	X	
Cash payments to suppliers	(X)	
Cash paid to and on behalf of employees	(X)	
Other cash payments	(X)	
Net cash inflow/(outflow) from operating activities		X or (X)
Returns on investments and servicing of finance		
Interest received	X	
Interest paid	(X)	
Finance charge element of finance lease rental payments	(X)	
Dividends received from trade investments	X	
Dividends paid	(X)	
Net cash inflow/(outflow) from returns on investments and servicing of finance		X or (X)
Taxation		
UK corporation tax paid	(X)	
Overseas tax paid	(X)	
Tax paid		(X)
Investing activities		
Purchase of tangible fixed assets	(X)	
Purchase of subsidiary undertakings (net of cash and cash equivalents acquired)	(X)	
Sale of tangible fixed assets	X	
Sale of trade investment	X	
Net cash inflow/(outflow) from investing activities		X or (X)
Net cash inflow/(outflow) before financing		X or (X)
Financing		
Issue of ordinary share capital	X	
Issue of debentures	X	
Long-term unsecured loan raised	X	
Redemption of debentures	(X)	
Capital element of finance lease rental payments	(X)	
Net cash inflow/(outflow) from financing		X or (X)
Increase/(decrease) in cash and cash equivalents		X or (X)

Table 6.2 Cash flow statement notes format

Notes to the cash flow statement

1. *Reconciliation of operating profit to net cash inflow/(outflow) from operating activities*

	£000
Operating profit	X
Depreciation charges	X
Profit on sale of tangible fixed assets	(X)
Loss on sale of trade investments	X
Increase in stocks	(X)
Decrease in debtors	X
Decrease in creditors	(X)
Net cash inflow/(outflow) from operating activities	X or (X)

2. *Analysis of cash and cash equivalents during the year*

	£000
Balance at 1 January Year 5	X or (X)
Net cash inflow/(outflow) during the year	X or (X)
Balance at 31 December Year 5	X or (X)

3. *Analysis of the balances of cash and cash equivalents as shown in the balance sheet*

	Year 5 *£000*	*Year 4* *£000*	*Change in year* *£000*
Cash at bank and in hand	X	X	X or (X)
Short term investments	X	X	X or (X)
Bank overdrafts	(X)	(X)	X or (X)
	X or (X)	X or (X)	X or (X)

4. *Analysis of changes in financing during the year*

	Share capital *£000*	*Debentures* *£000*	*Unsecured loans* *£000*	*Finance lease obligations* *£000*
Balance at 1 January Year 5	X	X	X	X
Cash inflows/(outflows) from financing	X or (X)	X or (X)	X or (X)	X or (X)
Balance at 31 December Year 5	X	X	X	X

7 Accounting Aspects of Long-term Contracts

7.1 Introduction

Many companies in the building and construction industry are engaged in long-term contract work. In this context 'long-term' means that work under the contract from the start until completion spans more than one financial year. Contracts for the construction of an office block, for example, a new motorway or a reservoir would all fall into the long-term category. Long-term contracts give rise to a number of accounting problems, including the calculation of profits/losses on contracts, the amount of profit/loss to pass through the profit and loss account of a particular accounting period, and the valuation to be placed on contract work-in-progress. These and other points are the subject of this chapter.

7.2 Cost ascertainment

Most contractors are engaged on a number of long-term contracts simultaneously, each in differing stages of completion. For cost ascertainment purposes, each contract is assigned a unique code number under which all costs relating to the contract are collected. Costs encountered in long-term contracts include obvious ones such as labour and materials costs, and subcontractors' costs, together with other items including plant hire from other companies, depreciation of own plant used on site and production and functional administration overheads. (The nature, significance and accounting treatment of overheads is the subject matter of Chapters 9, 10 and 11 in Part II of this book.)

7.3 Profit/loss ascertainment

At regular intervals throughout the life of each contract it is necessary to arrive at an estimated figure of profit or loss on the completed contract. This is an essential operation at the end of each accounting period. At each of these ascertainments, the total estimated costs of each contract will consist of a varying mix of two elements – actual costs incurred to date and an estimate of further costs to completion. As each contract progresses, the first element forms a greater proportion of total cost with a consequential reduction in the second element. Further costs to completion include the costs of completing the remaining stages of the contract, including estimated future increases in wage rates and materials prices (to the extent that they are not covered by variation clauses); additionally, the cost of post-completion guarantee and rectification work must be estimated and included.

The combined figure of actual and estimated costs is compared with the contract price to deduce the estimated amount of profit or loss which the contract is likely to produce, thus:

	£000	£000
Contract price		X
less		
Actual costs incurred to date	X	
Estimated further costs to completion	X	
Estimated cost of guarantee and rectification work	X	
Provision for foreseeable losses	X	
Estimated total cost of contract		X
Estimated total profit/(loss) on contract		X or (X)

If the above calculation produces an estimated loss, it must be charged in full in the profit and loss account of the period in which the loss has been ascertained.

Where, however, an estimated profit is disclosed, the procedure is totally different. The accounting regulations are contained in a directive SSAP 9 (Stocks and Long-term Contracts) and in the Companies Act 1985. One effect of these regulations is that each contract has to be accounted for separately and that no apparent profit on any contract can be passed through the profit and loss account until the outcome of that contract can be assessed with reasonable certainty. This requires a decision to be made at senior management level, based on information and advice supplied by external (or the company's own) architects, surveyors and other technical staff. In practical terms, it is unlikely that the outcome of any contract could reasonably be foreseen until it is in the region of 70 per cent complete.

Accountants take this cautious approach because if, having taken too much profit into account too soon, it later transpires that a loss has been sustained on the contract, the company could find itself in serious financial difficulty. The reason for this is that on the strength of apparently high profits, the directors might declare and distribute a much higher dividend that they would otherwise have done. As the result of paying out non-existent profits, the company effectively would have seriously weakened its financial position.

When the calculation discloses that, on the basis of all the facts known at the time, a particular contract is likely to produce a profit and, at the same time, the outcome is regarded as being reasonably certain, the next question to arise is the amount of that estimated profit that can be passed through the profit and loss account of that accounting period.

Here, again, a conservative approach is adopted. Profit taken into account needs to reflect the proportion of work carried out by the accounting date, at the same time allowing for any inequalities of profitability in the various stages of the contract. As a further precaution, the profit to be recognised may be reduced even more by the application of a fraction – progress payments received as a proportion of those invoiced. Work carried out can be measured on one of two bases: *contract value* and *contract cost*; either method is acceptable under SSAP 9.

- *Contract value* – the following fraction is used:

  $$\frac{\text{Value of contract work completed}}{\text{Contract value (price)}}$$

  Figures in both the numerator and the denominator include the profit element.
- *Contract cost* – the following fraction is used:

  $$\frac{\text{Actual cost of work completed}}{\text{Estimated total contract cost}}$$

  Figures in both the numerator and the denominator exclude the profit element.
- *Other reduction factors*

  (i) $$\frac{\text{Progress payments received}}{\text{Progress payments invoiced}}$$

  The denominator includes payments actually received (shown as the numerator) plus amounts invoiced but not yet received and retention moneys.

  (ii) Other, as necessary, to reflect inequalities of profitability.

In any accounting period, the profit figure to which the above fractions are applied is the total estimated profit on the contract calculated as previously described. The figure of profit resulting from this calculation is the cumulative profit to be recognised to date. It is then a simple matter to arrive at the profit to be recognised in the current period by deducting the previous year's cumulative profit recognised from that of the current year. An example should clarify the points made so far:

Example

At 31 December Year 6, the following data was available in respect of one contract in course of completion by a construction company:

	£000
Contract value	8 000
Cumulative: to 31 December Year 6	
Costs incurred	4 000
Progress payments received	3 000
Progress payments invoiced	3 600
to 31 December Year 5	
Profit recognised	974
Other data at 31 December Year 6	
Estimated further costs	
To completion	1700
Future price increases not recoverable	100
Guarantee/rectification work	200

The outcome of this contract has been assumed with reasonable certainty.

Required

Calculate the amount of profit which could be recognised in the profit and loss account for the year ended 31 December Year 6.

Solution

	£000	£000
Contract value		8 000
Costs		
Actual to date	4 000	
Estimated to completion	1700	
Estimated price increases	100	
Estimated guarantee/rectification	200	
		6 000
Estimated profit		2 000

Two alternative formulae have been given for use in calculating the amount of profit to be recognised – one based on value (including profit), the other based on cost. In this example, the former cannot be used because, in the absence of a figure for work completed awaiting certification, the total value of work completed to date cannot be ascertained. Therefore, on this occasion, there is no alternative but to use the second formula:

$$\frac{\text{Actual costs incurred to date}}{\text{Total estimated contract cost}} = \frac{4\,000}{6\,000} = \frac{2}{3}$$

For prudence, a reduction factor could be applied, to reflect the fact that some of the amount invoiced has not yet been received; in fact, the retentions element will not be received for a long time.

$$\frac{\text{Progress payments received}}{\text{Progress payments invoiced}} = \frac{3\,000}{3\,600} = \frac{5}{6}$$

	£000
Total profit to be recognised to 31 December Year 6 [2/3 × 5/6 × 2 000]	1111
less Profit recognised to 31 December Year 5	974
Profit to be recognised in year Year 6	137

7.4 Profit/loss disclosure

Prior to the revision in 1988 of the accounting rules contained in SSAP 9, it was usual for companies to disclose contract profits/losses as a single line entry 'Profits/(losses) on long-term contracts' in the profit and loss account. Revised rules which have operated since then require that such profits or losses be brought into account as the difference between a figure of contract turnover and contract cost of sales. The effect of this is that where a profit is to be recognised, contract turnover is greater than cost of sales by the amount of profit, and vice versa in the case of a loss. Where a contract has not progressed sufficiently for any profit to be recognised, the turnover and cost of sales figures are identical, thus:

Table 7.1 Recognition of contract profits/(losses)

	Contract 1 *£000*	**Contract 2** *£000*	**Contract 3** *£000*
Contract turnover	1 057	1 336	756
Contract cost of sales	843	1 413	756
Contract gross profit/(loss)	214	(77)	–
	Profit recognised	*Loss recognised*	*No profit or loss recognised*

SSAP 9 acknowledges that different companies use various methods of arriving at a figure of contract turnover. For that reason it does not prescribe any particular method and leaves companies the freedom to adopt any suitable method. In reality, accountants use variations of one of two basic methods.

Method 1 To arrive at turnover by adding the profit to be recognised (calculated as shown in Section 7.3) to the cost of sales figure.

Method 2 To arrive at a figure of turnover, based on engineers'/architects' certificates, deemed to include the profit to be recognised (calculated as shown in Section 7.3). Cost of sales can then be deduced from the turnover, as being equal to the turnover minus the profit element.

In the event of a loss on a contract, the provision for loss forms part of the cost of sales figure together with actual costs incurred.

As has been seen already, contract profit recognised in an accounting period is the difference between the cumulative profit to the end of the current and previous accounting years. Similar principles apply in the ascertainment of contract turnover and cost of sales, both of which in the accounts of the current year are the difference between the figures of the current and previous year's cumulatives, as is illustrated below.

Table 7.2 Derivation of profit and loss account contract amounts

	Contract 4	**Contract 5**	**Contract 6**
	£000	*£000*	*£000*
Turnover			
Cumulative to 31 December Year 6	857	413	241
Cumulative to 31 December Year 5	724	367	198
For year to 31 December Year 6	133	46	43
Cost of sales			
Cumulative to 31 December Year 6	672	452	241
Cumulative to 31 December Year 5	581	394	198
For year to 31 December Year 6	91	58	43
Gross profit			
Cumulative to 31 December Year 6	185	(39)	–
Cumulative to 31 December Year 5	143	(27)	–
For year to 31 December Year 6	42	(12)	–

If these were the only activities of the company during the year Year 6, the profit and loss account would disclose aggregate figures, thus:

	£000
Turnover	222
Cost of sales	192
Gross profit	30

7.5 Balance sheet disclosures

Sums relating to long-term contracts have to appear, as appropriate, under one or more of the following headings:

Debtors
Amounts recoverable on contracts

Creditors
Payments on account

Stocks
Long-term contract balances

Provision for liabilities and charges
Provisions for losses on contracts

The meaning of each of these items will now be explained.

- *Debtors: amounts recoverable on contracts* is the net excess, on a cumulative basis, of contract turnover over payments on account (as now defined). For example:

	Contract 71
	£000
Cumulative to 31 December Year 6	
Turnover	1596
Payments on account	1500
Amounts recoverable on contracts at 31 December Year 6	96

- *Creditors: payments on account* is the net excess, on a cumulative basis, of contract payments on account over turnover. For example:

	Contract 72
	£000
Cumulative to 31 December Year 6	
Turnover	1 440
Payments on account	1 600
Payments on account at 31 December Year 6	(160)

If, however, long-term contract balances arise (as shown subsequently) payments on account can be set against them to the maximum extent available:

	Contract 72
	£000
Cumulative to 31 December Year 6	
Long-term contract balance	130
Transfer from payments on account	(130)
Long-term contract balance at 31 December Year 6	Nil

The effect on payments on account, as previously shown, is:

	Contract 72
	£000
Payments on account	160
Transfer against long-term contract balance	(130)
Payments on account at 31 December Year 6	30

- *Stocks: long-term contract balances* is the net excess, on a cumulative basis, of costs incurred on contracts over costs transferred to contract cost of sales. In its basic form it represents work-in-progress at cost. For example:

	Contract 73
	£000
Cumulative to 31 December Year 6	
Costs incurred on contract	1620
Costs transferred to cost of sales	1292
Long-term contract balance at 31 December Year 6	328

For disclosure purposes, long-term contract balances are reduced, as applicable, not only by excess payments on account as illustrated previously, but also by provisions for foreseeable losses. For example:

	Contract 73
	£000
Long-term contract balance (as above)	328
Provision for foreseeable loss	(270)
	58
Excess payments on account	(41)
Long-term contract balance at 31 December Year 6	17

- *Provisions for liabilities and charges: provisions for losses on contracts* is the excess, on a cumulative basis, of the provision for foreseeable losses over long-term contract balances. For example:

	Contract 74
	£000
Long-term contract balance	230
Provision of foreseeable loss	350
Provision for foreseeable loss at 31 December Year 6	(120)

The term 'payments on account' is misleading in two ways. First, it applies to payments by the contractee (client customer) to the contracting company, in whose hands they are receipts. Second, it includes amounts actually received together with amounts receivable – progress payments invoiced but not yet received and retentions – at the end of the accounting period.

Up to this point the circumstances of each of Contracts 71 to 74 used in the examples have been viewed in isolation. The various debit and credit balances arising would not be netted off against each other in the balance sheet because the netting off of revenues, expenses, assets and liabilities is expressly forbidden by the Companies Act 1985. Thus, the various balances attributable to the four contracts would appear in the balance sheet as now shown:

Balance Sheet (extract) as at 31 December Year 6

	£000
Current assets	
Stocks	
Long-term contract balances [Contract 73]	17
Debtors	
Amounts recoverable on contracts [Contract 71]	96
Creditors: amounts falling due in more than one year	
Payments on account [Contract 72]	30
Provisions for liabilities and charges	
Provision for foreseeable losses [Contract 74]	120

Additionally, because the gross figures for long term contract balances have been reduced by excess payments on account, SSAP 9 requires that a note should be annexed to the balance sheet to clarify the position for users of the financial information.

Note:

	£000
Long-term contract balances (less foreseeable losses) [Contracts 72 and 73] [130 + 58]	188
less	
Applicable payments on account [Contracts 72 and 73] [130 + 41]	171
Long-term contract balances [included under Stocks in Balance Sheet]	17

8 Evaluation and Interpretation of Financial Statements

8.1 Introduction

The financial statements that it is obligatory for a company to produce and publish within the Annual Report and Accounts are those encountered and explained in earlier chapters – the profit and loss account, balance sheet and cash flow statement. As far as the company is concerned, the production of these statements is an end it itself. For all parties, except for the directors and senior executives of the company, the financial statements are the only accessible items of financial data and information about the company's performance and financial position apart from statistics and commentaries which may appear in the financial press.

The various parties have differing reasons for their interest, as the result of which they are interested in diverse aspects of the company. It will therefore be helpful to identify the most common actual and potential users of company published accounts, the aspects in which they are interested and the reasons for their interest.

8.2 Users of published accounts of companies

Apart from the directors and senior executives who, as was stated above, have privileged access to financial data and information which is not published, the main groups of users of published financial data publicly available are:

- shareholders;
- employees;
- lenders;
- suppliers;
- customers;
- government departments;
- competitors.

The respective interests of each of the above parties will now be identified, together with the extent of their interests and the reasons for them.

- *Shareholders*, both institutional and individual, invest in a company in the expectation of future income, in the form of a stream of dividends, and of growth in the value of the investment. They are therefore interested in those aspects which indicate the extent to which their objectives are likely to be achieved. Of particular interest, therefore, is data concerned with earnings, dividend payments, book value of assets, share prices and market capitalisation (that is, the number of issued shares multiplied by their market value per share).

- *Employees*, and their trade union representatives, are interested in those aspects of a company's performance, such as financial stability, which indicate their long-term employment prospects and security. The prosperity of the company is a major factor in determining the size and timing of pay claims, consequently employees are interested in the company's earnings and earnings potential.
- *Lenders* need to be assured that the company can earn sufficient profits to meet the interest charges. Of even greater importance is the prerequisite that the company will be able to pay these interest charges and repay the principal sums by instalments or in lump sums according to the terms of the loans.
- *Suppliers'* interests are very similar to those of lenders. Both parties are creditors of the company. While lenders can be long- or short-term creditors, suppliers, in the main, are short-term creditors. Their immediate concern is the assurance that their invoices will be paid on the due dates. If a large proportion of their supplies are taken up by a company, the supplier will be very anxious to examine the long-term prospects and financial stability of that company, because their two interests will be inextricably linked.
- *Customers* have interests which coincide in many respects with those of suppliers. In a situation where a customer relies upon a company to supply a major quantity of its purchases, the failure of the company would have a detrimental effect on the financial stability of the customer. In order to avoid or minimise disruption of its own operations, the customer would be forced to seek out alternative sources of supply at short notice, and usually on disadvantageous terms. Thus customers, like suppliers, are interested in the long-term prospects and financial stability of the company.
- *Government departments* use the financial data of companies for a variety of purposes. The Department of Trade and Industry (DTI) uses them for compiling statistical bulletins and surveys, and for production census purposes. HM Customs and Excise use the figures for calculating VAT collectible from or refundable to companies. Income tax and corporation tax liabilities of companies are the concern of the Inland Revenue.
- *Competitors* are interested for one or other of two main reasons. First, they are interested in comparing their own financial performance and position with those of other companies to identify their own strengths

and weaknesses. Second, if they have an acquisitive policy, they are on the lookout for other companies for which they could feasibly make a take-over bid.

8.3 Evaluation methods

The basic principle of evaluation is the recognition that there are a number of financial relationships within the profit and loss account, the balance sheet and the cash flow statement, not only as separate, individual financial reports but also between one another.

Assets/net assets shown in the balance sheet generate the sales (turnover) disclosed in the profit and loss account; within the profit and loss account gross profit and net profit originate from sales. From the foregoing facts it follows that there must also be a relationship between net profit in the profit and loss account and assets/net assets or capital employed in the balance sheet, the common link being sales; this relationship is termed *profitability*.

Relationships of the type described above are usually expressed as ratios. The term *ratio* is used broadly to include a variety of forms – proportions, number of times, percentages and other forms – depending upon the item concerned. The calculation of ratios as an end in itself would have limited value. Real benefits arise from ratios only when they are compared with other ratios and the differences between the various sets are measured and investigated with a view to executive action to effect an improvement.

8.4 Ratio comparisons

Companies compare the ratios for the current financial period with a variety of different ratios. The nature and purpose of commonly encountered comparisons are listed below.

- *Comparison with previous period(s)* discloses improvements and/or deteriorations from the previous period. When this is extended to the financial results of the last five or ten periods it is relatively easy to identify trends. A shortcoming of this comparison is that it indicates differences in what has *actually* taken place not with what *should have* taken place.
- *Comparison with forecast* partially overcomes the drawbacks of the previous comparison in that ratios calculated on actual results are compared with those based on an estimate made some time previously of what would be likely to happen in the current period. Differences between the two sets of ratios show the effects of divergence from forecast.

- *Comparison with budget* is, from a financial control viewpoint, the most useful of all comparisons. Ratios calculated on the actual results of the current period are compared with their counterparts in the budget – the predetermined planned target. Budgetary control is the subject matter of Chapter 15, but at this juncture it is sufficient to state that significant discrepancies between actual and budgeted figures have to be taken very seriously by management because they indicate that plans are not being realised. This in turn leads to an investigation into reasons for the discrepancies and the formulation of plans to bring the company back in line with the target.
- *Comparison with competitors* is a worthwhile exercise because it enables a company to compare its own financial relationships with those within the same industry or sector. There are two main sources of this information. First, the published Annual Report and Accounts can be obtained for any company from either the company itself or, failing that, from the Registrar of Companies on payment of search and extract fees. Second, and more usefully, businesses within a particular industry often belong to a trade association. The trade association receives the financial data from each of its member businesses, collates it in standardised form, and circulates the results of the individual businesses to each of its members. In the interests of confidentiality, the results of individual businesses are not identified by name. However, each recipient company is able to recognise its own results and is thus able to see how it is performing in relation to other businesses in the same industry. The validity of the comparisons is enhanced because ratios eliminate differences in absolute figures arising from differences in scale of operations, by restating figures to a common base. Thus the performance of a small company with a net profit of £100 000 on a capital employed of £1 000 000 can be compared with that of a larger company producing a profit of £1 500 000 from a capital employed of £30 000 000. Their respective profitability, as described in Section 8.3 above, is 10 per cent and 5 per cent in ratio terms.

8.5 Ratio groupings

Calculation of ratios falls into four recognisable groups, which are explained in Sections 8.6 to 8.9 to follow, namely, those concerned with *profitability; solvency and liquidity; gearing;* and *investment.* In practice there are numerous ratios, of varying degrees of sophistication. This chapter will be confined to the consideration only of those most commonly employed.

Throughout ratio analysis there is a problem of definitions. It is an unfortunate fact that there are no universally agreed definitions of the component elements of the ratios. Thus, when considering the ratios of different companies, allowance has to be made if the bases of calculation are not uniform. Profit and capital employed are two examples of this problem. For calculation purposes, profit can mean profit before tax, profit after tax, profit before interest and tax (also known as operating profit) and earnings (profit after tax attributable to equity shares) while capital employed can be defined variously as shareholders' funds plus long term creditors, fixed assets plus net current assets, and total net assets. The basis of each ratio used in this chapter will be stated at each point, but it must be remembered that there are these variations. Provided that the ratios being compared have been calculated, or adjusted, to the same basis, these differences lose their importance. Of greater significance is the fact that they are consistent over time and between each other.

Before proceeding to the formulae for and calculation of specific ratios, an illustrative set of final accounts in condensed form is given in Table 8.1 and 8.2.

Table 8.1 Illustrative profit and loss account

Bill Ding (Supplies) Ltd

Profit and loss accounts (summarised)
for the years ended 31 December

	Year 4		Year 5	
	£000	£000	£000	£000
Turnover		2 500		3 250
Cost of sales		1 650		2 100
Gross profit		850		1 150
Distribution costs	260		350	
Administrative expenses	170		240	
		430		590
Operating profit		420		560
Interest payable		70		150
Profit before tax		350		410
Taxation		90		100
Profit after tax		260		310
Dividends proposed		80		90
Retained profit for the year		180		220

Table 8.2 Illustrative balance sheet

Bill Ding (Supplies) Ltd
Balance sheet (summarised) as at 31 December

	Year 4		*Year 5*	
	£000	*£000*	*£000*	*£000*
Fixed assets				
Tangible (various)		2 080		2 970
Current assets				
Stock	427		709	
Trade debtors	270		420	
Bank and cash	70		111	
	767		1 240	
Creditors due within one year				
Trade creditors	215		400	
Other creditors (incl. tax and dividends)	207		265	
	422		665	
Net current assets		345		575
Total assets less current liabilities		2 425		3 545
Creditors due in more than one year				
Loans repayable 20×7		1 000		1 900
Share capital and reserves				
Called up share capital	500		500	
Profit and loss	925		1 145	
Shareholders' funds		1 425		1 645
Capital employed		2 425		3 545

8.6 Profitability ratios

The connection between assets/net assets/capital employed, sales and net profit has already been mentioned in Section 8.3 and described as *profitability*. Various ratios can be used to measure this relationship but the main ones are:

(a) gross profit to sales percentage;
(b) operating profit to sales percentage;
(c) sales to net assets (capital employed);
(d) operating profit to net assets (capital employed) percentage; and
(e) net profit (after tax) to shareholders' funds.

Each of these is now explained using figures from the summarised accounts of Bill Ding (Supplies) Ltd: in Tables 8.1 and 8.2.

- *Gross profit to sales percentage* measures the profit component of the sales figure after allowing for the input cost of the items sold.

Formula: $$\frac{\text{gross profit}}{\text{sales (turnover)}} \times 100$$

Year 4	Year 5
$\frac{850}{2\,500} \times \frac{100}{1} = 34.0\%$	$\frac{1\,150}{3\,250} \times \frac{100}{1} = 35.3\%$

Changes in the gross profit ratio can reflect a number of factors, including changes in the mix of sales of different products with different individual gross profit margins; absorption of higher prices charged by suppliers on bought-in goods and materials; pricing strategies; and changes in market conditions.

- *Operating profit to sales percentage* measures the profit component of the sales figure after allowing for all costs concerned with operating and trading, but disregarding financing items.

Formula: $$\frac{\text{operating/trading profit}}{\text{sales (turnover)}} \times \frac{100}{1}$$

Year 4	Year 5
$\frac{420}{2\,500} \times \frac{100}{1} = 16.8\%$	$\frac{560}{3\,250} \times \frac{100}{1} = 17.2\%$

Improvements and deteriorations in the operating efficiency of the business are highlighted by movements in this ratio.

- *Sales to net assets (capital employed) ratio* measures the efficiency with which the company's net assets have been utilised. Differences in definition were noted in Section 8.5, but for this illustration, net assets are taken to mean fixed assets plus net current assets.

Formula: $$\frac{\text{sales (turnover)}}{\text{net assets (capital employed)}}$$

Year 4	Year 5
$\frac{2\,500}{2\,425} = 1.03$	$\frac{3\,250}{3\,545} = 0.92$

An alternative way of viewing the above result is that it shows the amount of sales generated by each £1 of net assets. Frequently the analysis is taken a stage further by limiting the figure in the denominator to fixed assets. The resultant figures then show the success, or lack of it, of capital expenditure investment programmes.

Formula: $$\frac{\text{sales (turnover)}}{\text{fixed assets}}$$

Year 4	Year 5
$\frac{2\,500}{2\,080} = 1.20$	$\frac{3\,250}{2\,970} = 1.09$

Care has to be taken when interpreting calculations involving fixed assets and net assets. Fixed assets are normally included at their written down value, which reduces with each year's depreciation charge. Thus, supposing the sales and fixed assets of the consecutive periods were identical, the sales to assets ratios of the second period would be higher than those of the first period, not by reason of improved efficiency but solely because the denominator used in the second period would be smaller, by the amount of that period's depreciation, than the denominator used in the first period's calculation. The improvement would therefore be illusory. Further calculations may be employed to counteract this phenomenon but are outside the scope of this book.

- *Operating profit to net assets (capital employed) percentage* is usually referred to as the return on capital employed (ROCE) and measures overall profitability. It links the profit with the resources employed in generating it.

Formula: $$\frac{\text{operating/trading profit}}{\text{net assets (capital employed)}} \times \frac{100}{1}$$

Year 4	Year 5
$\frac{420}{2\,425} \times \frac{100}{1} = 17.32\%$	$\frac{560}{3\,545} \times \frac{100}{1} = 15.80\%$

Return on capital employed is governed by a combination of the two relationships dealt with previously – operating profit to sales and sales to net assets. This can be easily demonstrated by referring to the figures previously calculated.

Formula: operating profit to sales percentage × sales/net assets ratio

Year 4	Year 5
$16.8\% \times 1.03 = 17.3\%$	$17.2\% \times 0.92 = 15.8\%$

From this it can be seen that a company can improve its overall profitability by improving either or both of its component elements. Similarly, a deterioration in either or both of these elements will have an adverse effect on overall profitability. A third possibility is that an improvement in one element can be counteracted by a deterioration in the other, leaving overall profitability unchanged.

- *Net profit (after tax) to shareholders' funds percentage* is also known as return on shareholders' funds. It is usually related to ordinary (equity) shares. Thus, if preference shares are in issue, preference dividends are deducted from net profit after tax and preference share capital from shareholders' funds, leaving ordinary shareholders' funds and the profit after tax attributable to them.

Formula: $\frac{\text{net profit (after tax)}}{\text{shareholders' funds}} \times \frac{100}{1}$

Year 4	Year 5
$\frac{260}{1\,425} \times \frac{100}{1} = 18.25\%$	$\frac{310}{1\,645} \times \frac{100}{1} = 18.84\%$

This ratio indicates the effects of the return on capital employed on funds, which include share capital and undistributed profits invested by shareholders, after taking into account the impact of financing items and of taxation.

8.7 Solvency and liquidity ratios

While it is important for any business to maintain a satisfactory level of profit and profitability (as measured in Section 8.6), it is essential that it does so from a position of financial strength and soundness. There are numerous instances of companies in all sectors of industry and commerce which, though highly profitable, have nevertheless failed financially and have been forced into receivership or liquidation. High profitability is no guarantee of survival.

Ratios in this group measure solvency – the extent to which the assets of a business exceeds its external obligations (liabilities); and liquidity – the degree of availability of those assets in a form readily available to settle those obligations as they fall due.

The main ratios used to measure these relationships are:

(a) current ratio;
(b) acid test ratio;
(c) stock turnover ratio;
(d) debtors' collection period;
(e) creditors' payment period; and
(f) operating cash cycle.

Each of these is now explained, using figures from the summarised accounts of Bill Ding (Supplies) Ltd from Section 8.5.

- *Current ratio* is a measure of the total current assets in relation to total current liabilities, that is, creditors falling due within one year. It is merely a general indicator acting as a starting point for the more detailed calculations in ratios (b) to (f) listed earlier.

Formula: $$\frac{\text{current assets}}{\text{current liabilities}} : 1$$

Year 4	Year 5
$\frac{767}{422} : 1 = 1.82 : 1$	$\frac{1\,240}{665} : 1 = 1.86 : 1$

A figure of greater than 1:1 shows that current assets exceed current liabilities, but a ratio of less than 1:1 indicates that current liabilities are greater than the current assets available for their settlement. Very often this means that the business cannot pay its way and is therefore insolvent.

- *Acid test ratio* is a more refined measure than the current ratio, because it takes into the numerator only those items which are actually in cash (and therefore immediately available to meet external liabilities) and those items, including debtors and marketable securities, which can be converted into cash at relatively short notice. By definition, therefore, stocks of all kinds, raw materials, work-in-progress (including long-term contract work-in-progress – see Chapter 7, Section 7.5), consumables, and finished goods awaiting sale are all excluded from this calculation. It is also known by the alternative names of *quick assets ratio* and *liquid assets ratio.*

Formula: $$\frac{\text{current assets (excluding stocks)}}{\text{current liabilities}} : 1$$

Year 4	Year 5
$\frac{(270 + 70)}{422} : 1 = 0.81 : 1$	$\frac{(420 + 111)}{665} : 1 = 0.80 : 1$

A figure of less than 1:1 for this ratio does not automatically mean that the company is insolvent. The reason for stating this is that it has been calculated using figures of current assets immediately available as the numerator, but the denominator contains items including taxation and dividends which, though due for settlement in the foreseeable future, do not have to be met immediately, In other words, this calculation does not reflect timings.

Some successful companies regularly manage on seemingly impossibly low acid test ratios. The secret of their survival is the fact that all or most of their business is conducted on a cash basis. Retail supermarkets fall into this category.

- *Stock turnover ratio* shows the number of times in the period that the stock is turned over, that is, used and/or sold and replenished.

Formula: $$\frac{\text{cost of goods sold}}{\text{average stock}}$$

Year 4	Year 5
$\frac{1\,650}{427} = 3.86$ times	$\frac{2\,100}{709} = 2.96$ times

An alternative way of viewing this is to calculate the number of days, on average, for which items of stock are held.

Formula: $$\frac{\text{average stock}}{\text{cost of goods sold}} \times \frac{365}{1}$$

Year 4	Year 5
$\frac{427}{1\,650} \times \frac{365}{1} = 94.46$ days	$\frac{709}{2\,100} \times \frac{365}{1} = 123.23$ days

The relationship between the two sets of figures is obvious. They are merely viewing stock turnover from different perspectives. A velocity of turnover in Year 4 of 3.86 times is just another way of stating that, on average, stock is held for a period of 94.46 days before being used or sold.

Separate calculations would be made for the different categories of stock – raw materials, work-in-progress and finished goods – where applicable.

A low velocity figure can indicate a number of factors detrimental to the business, including too high a stock level for the volume of sales (overstocking), and the presence of slow-moving stocks which could become worthless through obsolescence and/or deterioration.

- *Debtors' collection period* is the average length of time taken by customers from the date of the invoice for goods sold to them on credit, to the date of payment by them. It is usually measured in days, but can also be calculated in weeks, months or number of times in the period.

Formula: $$\frac{\text{trade debtors}}{\text{credit sales}} \times \frac{365}{1}$$

Year 4

$$\frac{270}{2\,500} \times \frac{365}{1} = 39.42 \text{ days}$$

Year 5

$$\frac{420}{3\,250} \times \frac{365}{1} = 47.17 \text{ days}$$

In reality, figures taken from the accounts have to be adjusted before being used in this formula because the debtors in the numerator include (VAT), whereas the sales figure in the denominator excludes it. The two figures have to be adjusted to the same basis.

This ratio measures the effectiveness of the credit control of the business. It enables management to see whether customers are paying the sums owed within the stipulated period of credit. Frequently this measure is used in conjunction with an 'ageing' schedule which analyses debtors according to the length of time for which their debts have been outstanding – less than one month, one month, two months, three months and more than three months.

By these means, slow paying and potentially defaulting customers can be identified and action taken to secure payment and/or stop the supply of further goods and services.

- *Creditors' payment period* is the average length of time taken by the business from the date of receipt of an invoice for goods and services bought from suppliers on credit, to the date of payment to them. It is usually measured in days, but can also be calculated in weeks, months or number of times in the period.

Formula: $$\frac{\text{trade creditors}}{\text{credit purchases (or cost of sales)}} \times \frac{365}{1}$$

Year 4	Year 5
$\frac{215}{1\,650} \times \frac{365}{1} = 47.56$ days	$\frac{400}{2\,100} \times \frac{365}{1} = 69.52$ days

A similar difficulty with VAT arises with this calculation as was noted with the debtors' collection figure. The denominator has to be adjusted to a VAT-inclusive basis.

This ratio alerts management to the degree of compliance with the terms of credit allowed by their suppliers. If the business is in breach of them, supplies could be stopped without warning and with disastrous consequences for the business.

- *Operating cash cycle* measures the interval of time between payment for goods and the receipt of cash for them. The simple diagram shown in Figure 8.1 explains the concept.

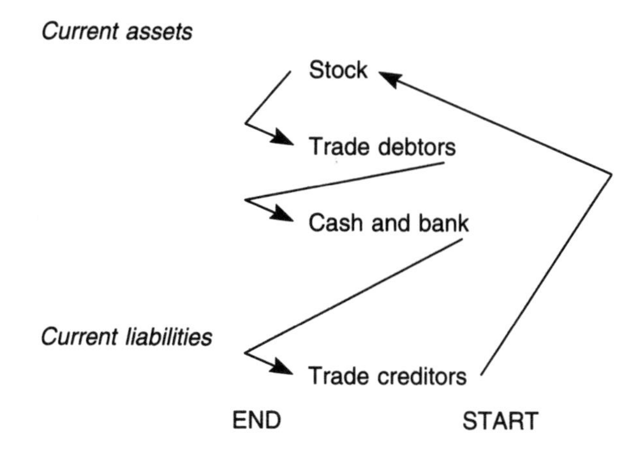

Figure 8.1 Operating current assets and liabilities

Goods are obtained on credit from suppliers. These are then sold, again on credit, to customers. The customers settle their accounts, and the cash goes into the bank, from which source the original suppliers are paid. This is a continuous cycle, the net length of which is measured by this calculation.

Formula: stock turnover days plus debtors' collection period minus creditors' payment period

	Year 4 *Days*	*Year 5* *Days*
Stock turnover	94.46	123.23
Debtors' collection	39.42	47.17
	133.88	170.40
Creditors' payment	(47.56)	(69.52)
Net operating cash cycle	86.32	100.88

In the above calculations all items of a non-operating nature would be eliminated. Thus figures for fixed asset debtors and creditors, and loan creditors, for example, would be excluded as being not relevant. The resultant figure is the number of days' cash supply being used to finance day-to-day trading operations.

8.8 Gearing ratios

As has been seen already in Chapter 2, companies are usually financed by a combination of shareholders' funds and external borrowings. Shareholders' funds consist of share capital and reserves, while external borrowings can take a variety of forms, including debentures, long term loans and convertible loan stock. The relationship between these external borrowings and the (internal) shareholders' funds is termed *gearing* or alternatively, *leverage*. Preference share capital is usually classed alongside borrowings.

Formula:

$$\frac{\text{long-term external borrowings plus preference capital}}{\text{long-term external borrowings plus shareholders' funds}} \times \frac{100}{1}$$

Year 4

$$\frac{1\,000}{2\,425} \times \frac{100}{1} = 41.23\%$$

Year 5

$$\frac{1\,900}{3\,545} \times \frac{100}{1} = 53.6\%$$

(There are other less commonly encountered versions of this formula which are not illustrated here.)

This ratio is an indicator of risk. A high percentage indicates a high risk to the lenders in that in times of poor liquidity, there may be insufficient cash available to meet interest payments and/or loan repayments. The dividing line between high and low risk is generally taken to be between 30 per cent and 40 per cent.

Another aspect of gearing is that, in conjunction with profitability ratios, it can be seen whether the interest cost of the borrowings is greater or lesser than the return on capital employed. Obviously, it is disadvantageous for the cost of borrowed funds to exceed the return produced by their investment.

A further calculation associated with borrowings is that of *interest cover* – the calculation of the number of times that interest payable has been covered by profit before tax and interest.

Formula: $$\frac{\text{profit before tax and interest}}{\text{interest payable}}$$

Year 4	Year 5
$\frac{420}{70} = 6$ times	$\frac{560}{150} = 3.7$ times

As a general rule, it would be undesirable for a company's cover to fall below 2. By reason of interest payable being a charge in arriving at earnings (profits available to ordinary shareholders), a low cover would mean that equity shareholders would receive either no dividend pay-out, or only a low one, unless accumulated distributable reserves could be tapped.

Another implication of gearing is that lending institutions, such as banks, are not prepared to make advances when a company has a high gearing percentage, it being considered an unacceptable risk.

8.9 Investment ratios

As has been seen already in Section 8.2, shareholders, both actual and potential, are interested in the financial aspects of a business. Of special relevance are a group of ratios labelled *investment ratios,* all of which are concerned with the ability of the business to pay dividends.

The main ratios in this category are:

(a) earnings per share (EPS);
(b) dividend per share (DPS);
(c) dividend cover;
(d) earnings yield;
(e) dividend yield; and
(f) price/earnings ratio (P/E).

For the purposes of the illustrations which follow, the called-up share capital of Bill Ding (Supplies) Ltd disclosed in the company's balance sheet in Section 8.5 at a figure of £500 000, consists of 2 000 000 ordinary shares at a nominal value of 25p per share.

- *Earnings per share (EPS)* is the amount of earnings attributable to each ordinary share in issue and qualifying for a dividend payout during the period. The term 'earnings' means the profit after deducting taxation charges and preference dividends; in other words, it is that part of the profit (after tax) attributable to ordinary shareholders. Earnings per share is always expressed in pence (and decimals of a penny), and never in pounds, or decimals of a pound, sterling.

Formula: $\dfrac{\text{earnings (in pence)}}{\text{issued ordinary shares}}$

Year 4	Year 5
$\dfrac{260 \times 100}{2\,000} = 13.0\,\text{p}$	$\dfrac{310 \times 100}{2\,000} = 15.5\,\text{p}$

The resultant figures are the earnings attributable (in this instance) to each ordinary 25 pence share as the result of the year's activities.

- *Dividend per share (DPS)* shows how much of the earnings per share is in fact paid out. By implication, the difference in amount between earnings per share and dividend per share represents the amount of retained, or ploughed- back, profit per share.

Formula: $\dfrac{\text{ordinary share dividend (in pence)}}{\text{issued ordinary shares}}$

Year 4	Year 5
$\dfrac{80 \times 100}{2\,000} = 4.0\,\text{p}$	$\dfrac{90 \times 100}{2\,000} = 4.5\,\text{p}$

Under normal circumstances, the typical dividend per share figure is in the region of one-third of the earnings per share figure.

- *Dividend cover* serves a purpose for ordinary shareholders similar to that of the interest cover calculation described in Section 8.8 for lenders, in that it shows the degree to which the ordinary dividend is covered by the available profit after tax.

Formula: $\dfrac{\text{net profit after tax minus preference dividends}}{\text{ordinary share dividend}}$

Year 4	Year 5
$\dfrac{260}{80} = 3.3\text{ times}$	$\dfrac{310}{90} = 3.4\text{ times}$

This calculation is another facet of the dividend/earnings relationship, showing the profit retention policy of the business. The higher the number, the greater is the proportion of profit being retained.

- *Earnings yield* relates the earnings per share to the cost to the investor of acquiring the entitlement to those earnings, by using the market price, not the nominal value, per share. The result is expressed as a percentage.

Formula: $$\frac{\text{earnings per share}}{\text{market price per share}} \times \frac{100}{1}$$

For the purpose of this and the dividend yield the price/earnings calculations which follow it is assumed that the market price of each 25 pence ordinary share of Bill Ding (Supplies) Ltd was 145 pence and 170 pence at 31 December Year 4 and Year 5 respectively. (This would be an internal price because, by reason of this being a private company, the shares could not be marketed publicly.)

Year 4	Year 5
$\frac{13.0}{145} \times \frac{100}{1} = 9.0\%$	$\frac{15.5}{170} \times \frac{100}{1} = 9.1\%$

The figure gives the earnings return for the year on the cost of the investment and, as such, enables comparisons to be made with the corresponding figures of other companies.

- *Dividend yield* relates the dividend per share to the cost to the investor of acquiring the entitlement to that dividend, by using the market price, not the nominal value, per share. The result is expressed as a percentage.

Formula: $$\frac{\text{dividend per share}}{\text{market price per share}} \times \frac{100}{1}$$

Year 4	Year 5
$\frac{4.0}{145} \times \frac{100}{1} = 2.8\%$	$\frac{4.5}{170} \times \frac{100}{1} = 2.6\%$

The figure shows the percentage return for the year received on the outlay, thereby enabling comparisons to be made with the corresponding figures of other companies and with other forms of investment.

In practice, a more sophisticated calculation is made, to recognise the fact that, in the hands of the recipient, the actual dividend received is deemed to be a greater amount by reason of the associated tax credit. This aspect, however, is beyond the scope of this book and will not be explored beyond this mention.

- *Price/earnings (P/E) ratio* is effectively the inverse of the earnings yield, except that it is expressed as a number and not as a percentage.

Formula: $\dfrac{\text{market price per share}}{\text{earnings per share}}$

Year 4	Year 5
$\dfrac{145}{13.0} = 11.2$ times	$\dfrac{170}{15.5} = 11.0$ times

This figure is used extensively by investment analysts as a comparator of performance between companies. One advantage is that, being an index number, it can be used for companies operating in different currencies without losing its validity. The financial pages of newspapers use this figure of P/E in preference to its counterpart, earnings yield.

The market price per share used as the numerator in the P/E calculation reflects various factors, but notably the expectation of future earnings. A high P/E ratio is usually interpreted as indicating a strong confidence in the company's ability to generate earnings in forthcoming years. Similarly, a low figure is indicative of a lack of such confidence. It is not possible to be too specific, but a figure of 10 or above is relatively high, and below 10 is relatively low.

8.10 Financial information in the daily papers

National newspapers, notably the *Financial Times*, the *Times* and the *Daily Telegraph*, devote several pages to company share matters each day, to keep investors informed.

Company shares which have a Stock Exchange listing (that is, a *quotation*) are marketed at two difference prices: the *offer price* and the *bid price*. An investor buys shares at the offer price, which is higher than the bid price obtainable if those same shares were being sold. The difference between the two prices is known as the *spread* and in effect this represents the profit made by the intermediaries through whom the transactions are conducted. For shares which are being traded, the offer and bid prices change constantly throughout the working day. Figures quoted in the financial columns of newspapers are the middle market prices at close of business on the previous working day. Middle market means the average of the offer and bid prices.

The typical layout of information published in the newspapers is as set out in Table 8.3.

Table 8.3 Newspaper share data

High (pence)	*Low (pence)*	*Company*	*Price (pence)*	*+/− (pence)*	*Net div (pence)*	*Yield %*	*P/E*
242	136	XYZ	170	+2	6	4.4	15.7

Explanation of columns in Table 8.3 is as follows. The first two columns show the highest and lowest middle market prices at which XYZ's shares have been traded in the past year. The price column gives the middle price at close of business of the previous working day, that is, in respect of Friday for Saturday and Monday newspapers, Monday for Tuesday newspapers. The change, increase (+) or decrease (−), from the opening price is recorded in the +/− column. Where appropriate, the dividend per share paid to shareholders is shown in the next column with the yield percentage, calculated on a gross (tax credit inclusive) basis, next to it. The final column supplies the price/earnings ratio. As was stated in Section 8.9, a high P/E ratio usually indicates that the stock market has a high degree of confidence in the company's profit-making ability. If, however, the market analysts are wrong, as they sometimes are, it can indicate that the company is overvalued.

Once a week, the newspapers disclose what is termed the market capitalisation of each company's shares. On the basis of the figures in Table 8.3, if XYZ had 11 000 000 shares in issue, the market capitalisation would be £18 700 000 (11 000 000 shares at 170 pence each); if the nominal value of these shares is £1 per share, the total nominal value would be £11 000 000. Thus it can be seen that market capitalisation simply means the valuation, at current market price, of a company's called-up share capital.

8.11 Validity of ratio analysis

Ratio analysis is a useful tool for users of accounting statements. In particular, comparison of ratio figures of a business over a number of time periods enables trends to be identified.

Users must, however, be aware of the limitations of ratio analysis. When a ratio is calculated to several decimal places, there is a danger that it is invested with a spurious accuracy. It must be realised that, while the calculation is precisely correct, it is the result of figures which can be of dubious validity. For example, if a stock turnover figure is calculated on the basis of the average of opening and closing stocks, the result would be totally misleading if it were the company's policy to run down stock levels at year ends for balance sheet 'window dressing' purposes; for the remainder of the year the company might have high stock levels which would not have been reflected in the ratio.

Conversely, a low stock turnover ratio might be the result of a high year-end stock level preparatory to a marketing campaign in the early months of the following year.

Especial caution must be exercised when comparing the ratio statistics of different businesses. The reason is that businesses have different policies for, say, stock valuation and/or depreciation, or different modes of operating – one business may have bought its fixed assets while the others hold theirs under finance or operating lease contracts; each of these circumstances would affect the ratio calculations. Ratios, then, are useful as general indicators. Their great importance is that they highlight the need for further, more detailed, investigation.

PART II

Management Planning and Control Practices

Part I was concerned mainly with the conventions of preparing and presenting figures relating to those financial aspects of a business contained in the Annual Report and Accounts, which is available to individuals outside the business and to other entities.

Part II deals with those financial matters with which business management is involved in planning and controlling operations. Unlike the subject matter of Part I, the financial data produced for the various levels of management is not accessible to any outside individual or entity.

9 Cost Classification

9.1 Introduction

Cost ascertainment is essential for any business, whether in the manufacturing, processing, building or surveying sectors, or whether it provides a service such as that of a consulting engineer, architect, quantity surveyor or estates surveyor. The reason for this is fairly obvious. The costs of producing a product or service need to be known so that the selling price may be fixed at a level which will produce a profit. As has been seen in Part I, businesses need profits in order to survive. There are, however, occasions where a business will sell at below cost, and therefore make a loss on a particular project or product. This is almost invariably a short-term measure undertaken for a specific purpose, as, for example, to gain a foot-hold in a new market. Such a situation does not remove the need for cost ascertainment. It is essential to enable the full extent of the cost of the policy to be calculated. For purposes of ascertainment, costs are analysed in various ways and the total costs of jobs, processes, products, contracts and services can then be arrived at by a process of synthesis.

9.2 Basic classification

At the primary level, three elements of cost are recognised and collected: *labour costs*; *materials costs*; and *other costs*. Each of these will now be considered.

- *Labour costs* include payments for work done and services performed, but exclude payments by way of reimbursement of expenses incurred: for example, travelling expenses, which are an 'other cost'. Within the labour costs category are salaries and wages, including overtime premiums, production and other bonuses, incentive payments, standby allowances, allowances for working in adverse conditions, piecework and taskwork payments, and holiday and sick pay.
- *Materials costs* include the cost of materials used in, and in the course of, production. Apart from obvious items, costs of petrol, diesel oil, coke and other fuels fall under this heading.
- *Other costs* consist of all those costs which have not been covered by the two preceding headings. Included, therefore, are depreciation, repairs, maintenance, power, office expenses (printing, stationery, telephones), advertising, plant hire, rent and rates. Specifically included are payments to sub-contractors (for example, to plasterers, plumbers, electricians and painters), and payments to other professionals such as consultant engineers, surveyors, solicitors and accountants.

9.3 Secondary classification

Each of the three elements dealt with above – labour, materials and other costs – is further classified into *direct costs* and *indirect costs*.

- *Direct costs* are those cost elements which can be directly identified with the finished product. In a manufacturing situation direct costs would comprise the costs of raw materials, which, after passing through various operations and/or processes, form the finished product, together with the wages of the machine operatives and other costs directly incurred. In a building context, direct costs would include the cost of those materials which form part of the building to be constructed – for example, bricks, sand, cement, timber, roof tiles, the wages cost of the labour force carrying out the construction, and the cost of sub-contracting and other specialist services involved.

  Direct costs are collectively termed *prime costs*.
- *Indirect costs* are those elements of cost incurred to enable production to take place, or in course of production, but which do not themselves form part of the finished product. Examples of indirect labour costs on a building site are the wages and salaries of site office staff, supervisory and management staff, and site security staff. Materials which do not form part of the buildings but which are used during construction, for example, oils and greases for lubricating site machinery, fuel for graders, bulldozers, mobile cranes, portable generators and compressors, are classed as indirect costs. The majority of costs categorised in Section 9.2 as 'other costs' are, by nature, indirect.

  Indirect costs are collectively termed *overhead costs*, or simply *overheads*.

  The *total cost* of a product, job, contract or provision of a service is the aggregate of prime costs and overheads.

9.4 Analysis of overheads

For planning and control purposes, overheads are analysed according to the function which they serve (*functional analysis*) and the behaviour they exhibit (*behavioural analysis*).

- *Functional analysis* – three functions are recognised: *production*; *marketing (selling) and distribution*; and *administration*. Sometimes the combined marketing and distribution function is split into its two separate components.

  It is important to understand where the lines of demarcation (between the various functions) lie and to see examples of overhead costs which fall typically within each function. These are given in tabular form in Tables 9.1 and 9.2. These are typical examples of activities and their analysis.

Table 9.1 Functional analysis of overhead costs

Activity on which overheads are incurred	*Analysis of function*		
	Production	*Selling and distribution*	*Administration*
Securing orders		*	
Procurement of materials			
Ordering			*
Central storage	*		
Transport to site	*		
Site works from preliminaries to final completion	*		
Post-completion guarantee/rectification	*		
Sales office		*	
Show houses		*	
Head office			*

Table 9.2 Examples of specific overhead costs analysed according to function

Function		
Production	*Selling and distribution*	*Administration*
Central stores Wages Lighting Fork lift truck running expenses Insurance	**Sales office** Newspaper and magazine advertising Production of brochures Local authority rates Wages and salaries Travelling expenses	**Head office** Directors' fees and expenses Supplies office wages Heating and lighting Local authority rates Management salaries, national insurance and pension scheme contributions Finance department salaries Auditors' fees Legal and similar costs for land conveyancing and planning consents
Site works Site office Running expenses plant and equipment running costs, including repair and maintenance Site security	**Show houses** Display lighting Sales negotiators' expenses and commissions	

- *Behavioural analysis* – overheads behave in one of three ways: *fixed*; *variable*; and *semi-variable* (*or semi-fixed*).

This is an exceedingly important aspect of overhead analysis, the many implications of which will become apparent in most of the chapters which follow.

- *Fixed overheads* remain at a constant level over a wide span of activity. They are still incurred even if no activity is taking place at all, as in the case of a major breakdown or strike or in the aftermath of a disastrous fire. For that reason they are sometimes referred to as shut-down costs.

  Examples of fixed costs include depreciation of fixed assets, local authority rates, fixed wages and salaries, and insurance. When, however, a certain level of activity or output has been achieved, it can be exceeded only if additional fixed costs are incurred, such as those resulting from the acquisition of more plant and machinery, thereby increasing the depreciation charge, and/or the employment of extra fixed salaried supervisory staff.

  This phenomenon is illustrated in Figure 9.1, which shows fixed costs on a plateau until 75 per cent of maximum capacity has been reached, at which point further fixed costs are incurred, giving rise to a stepped effect and a higher plateau.

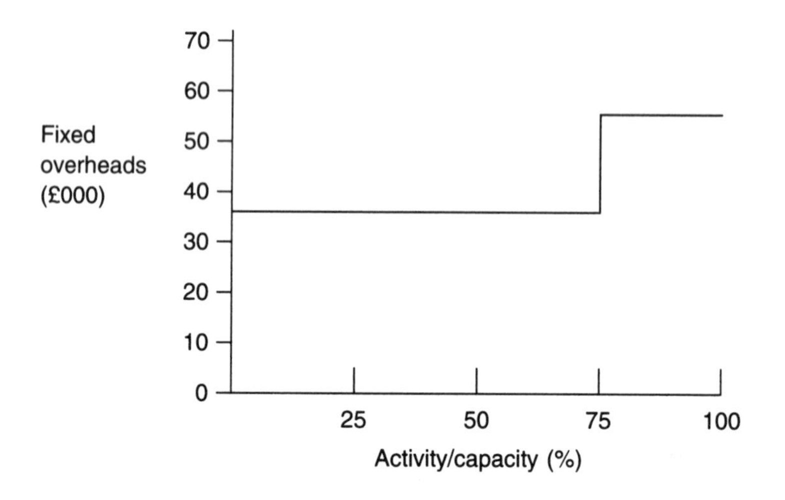

Figure 9.1 Fixed overheads behaviour

- *Variable overheads*, as their name implies, fluctuate with the level of activity. If no activity is taking place, say, over a statutory holiday period, mobile plant will not be using any fuel oil. Figure 9.2 shows the behaviour of variable overheads in diagram form. The variable overheads line has been shown as a straight line starting from zero. This implies that, in the absence of any activity, no variable overheads arise; but also that, when activity occurs, variable overheads increase proportionately. This latter presumption can be rebutted because, as activity increases, so does the scope for economies of scale. For example, a large order for consumable stores might qualify for more advantageous discount terms from the supplier. In that case the variable overhead cost would be in the form of a concave curve; this possibility has, however, been ignored for the purposes of the diagram.

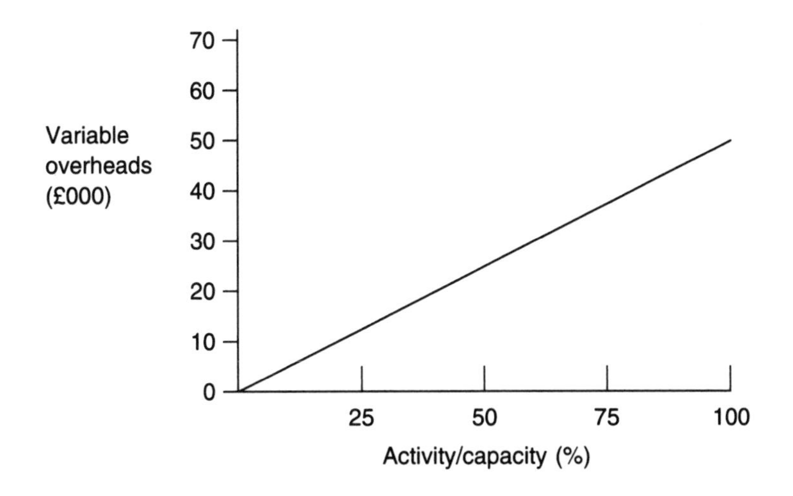

Figure 9.2 Variable overheads behaviour

- *Semi-variable overheads* are, as is to be expected from their label, a mixture of fixed and variable elements. An example of a semi-variable overhead cost is the power charge on the industrial tariff which consists of a fixed element, a standing monthly charge calculated on kilovolt amperes (KVA) of maximum demand, and a variable element – the charge for units consumed. A more easily understood and very common example is the charge for telephones. The fixed element is the line and equipment rental; the variable element is the number of (time) units used multiplied by the unit rate.

  Diagrammatic representation of semi-variable overhead costs is given in Figure 9.3. The semi-variable overhead line starts at a point above zero, thus reflecting the fixed overhead element.

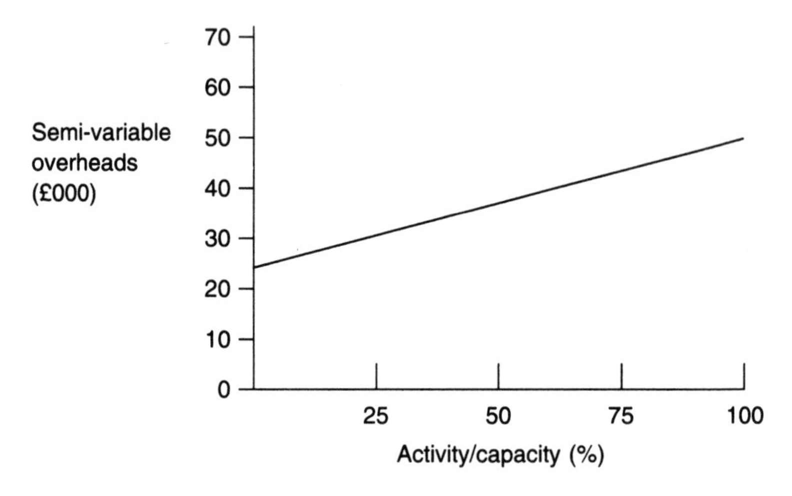

Figure 9.3 Semi-variable overheads behaviour

10 Overhead Attribution

10.1 Introduction

The need for arriving at the cost of any product was explained in Chapter 9. Total cost is built up in a series of layers, starting with prime cost (that is, with those costs which are directly referrable to the item being produced); in practical terms this means direct materials costs, direct labour costs and direct other costs. Overhead costs are then superimposed on the prime costs, on a functional basis (as illustrated in Tables 9.1 and 9.2 on p. 83 in a series of stages, thus:

	£
Prime cost	X
Production overheads	X
Production cost	X
Selling and distribution overheads	X
Administration overheads	X
Total cost	X

It is a comparatively easy matter for the total prime costs and total overhead costs, on a functional basis, to be estimated for a business as a whole. For reasons stated earlier, however, it is essential that the costs of individual jobs, products and contracts are calculated. This involves fairly complicated procedures to ensure that, as far as is possible, the figures of cost arrived at are a fair reflection of what has in fact been incurred. It is an accepted fact that in this area, absolute accuracy is impossible to achieve. The figures which emerge are accurate within reasonable degrees of tolerance, having been calculated on a logical basis, but based on estimates.

This chapter is concerned solely with the methods by which production overheads are attributed to production. Other overheads are dealt with in Chapter 11.

10.2 Production overhead attribution

Once the costs have been analysed into their primary and secondary classifications, as described in Chapter 9, the process of attribution to units of production can start. This is performed in stages, listed below, each of which involves specialised terminology which will be defined and explained in the detailed consideration of each stage which follows. The stages by which production overheads are attributed to products are:

Stage I	Attribution to production and service departments
Stage II	Attribution of service department overheads to production departments.
Stage III	Attribution of production department overheads to production cost centres.
Stage IV	Absorption of cost centre overheads by individual products.

Each of the above stages will now be considered separately.

10.3 Stage I: Attribution of production overheads to production and service departments

It is important at the outset to understand the distinction between production and service departments.

- A *production department* is one in which actual production operations and/or processes are carried out; in other words, a department through which a product passes during conversion from its raw state to its final completion. Easily understood examples can be obtained from manufacturing industry, where in a factory some or all of the following production departments can be found:

  preparation;
  moulding;
  machining;
  welding;
  assembly; and
  finishing.

- A *service department* is one which renders a service to production departments and to other service departments through which the products pass. Common examples of service departments are:

  power generation;
  stores;
  maintenance; and
  canteen.

The canteen, for instance, is used by operatives from the production departments, by storekeeping staff and by maintenance electricians and engineers.

Costs of operating the canteen are therefore shared out over the user departments.

Production overheads are attributed to production departments and to service departments by a process of allocation and apportionment.

- *Overhead allocation* is the allotment to departments of whole items of cost. The wages of the finishing department foreman can be allotted directly to that department; the cost of catering supplies can be allotted to the canteen.
- *Overhead apportionment* has to be carried out where an item of cost affects several departments generally. If premises are rented, the rental relates to the premises as a whole. Consequently, that proportion of the rental deemed to be attributable to the factory has to be split over the various production and service departments on an equitable basis. In this instance, relative floor areas occupied is usually regarded as the fairest basis for apportionment.

  Sometimes overheads which can be identified to specific departments, and which therefore are capable of being allotted, are nevertheless apportioned. This is the normal procedure where the administrative cost of keeping track of such overheads is uneconomic in relation to the amounts involved. As a matter of common sense the simpler, less costly, apportionment procedure is adopted.

An illustration of production overhead allocation and apportionment is given in Figure 10.1 (on p. 90) in diagram form for three production departments and two service departments. It can be seen from the diagram that the production overheads attributed to the five departments are a combination of allocated and apportioned items.

10.4 Bases of production overheads apportionment

Before proceeding to a consideration of Stage II, it is appropriate to examine the bases on which production overheads are commonly apportioned to departments. The principle followed is that the cost concerned is apportioned over departments in such a way as to reflect the factor(s) which caused the cost to be incurred in the first place. A few examples should make this clear. The canteen operating cost varies according to the demand for its services. It is usual to apportion the canteen cost on the basis of the number of personnel in each of the departments which it services, as a proportion of the total number of personnel employed in those departments. While not strictly an accurate basis, it is reasonable and therefore acceptable. Insurance premium costs can be apportioned on the basis of the relative insurable values of the items covered. Typical bases of apportionment for a selection of production overhead costs are given in Table 10.1, and a straightforward example of the application of the principles of apportionment is given in Table 10.2 using the departments employed in Figure 10.1.

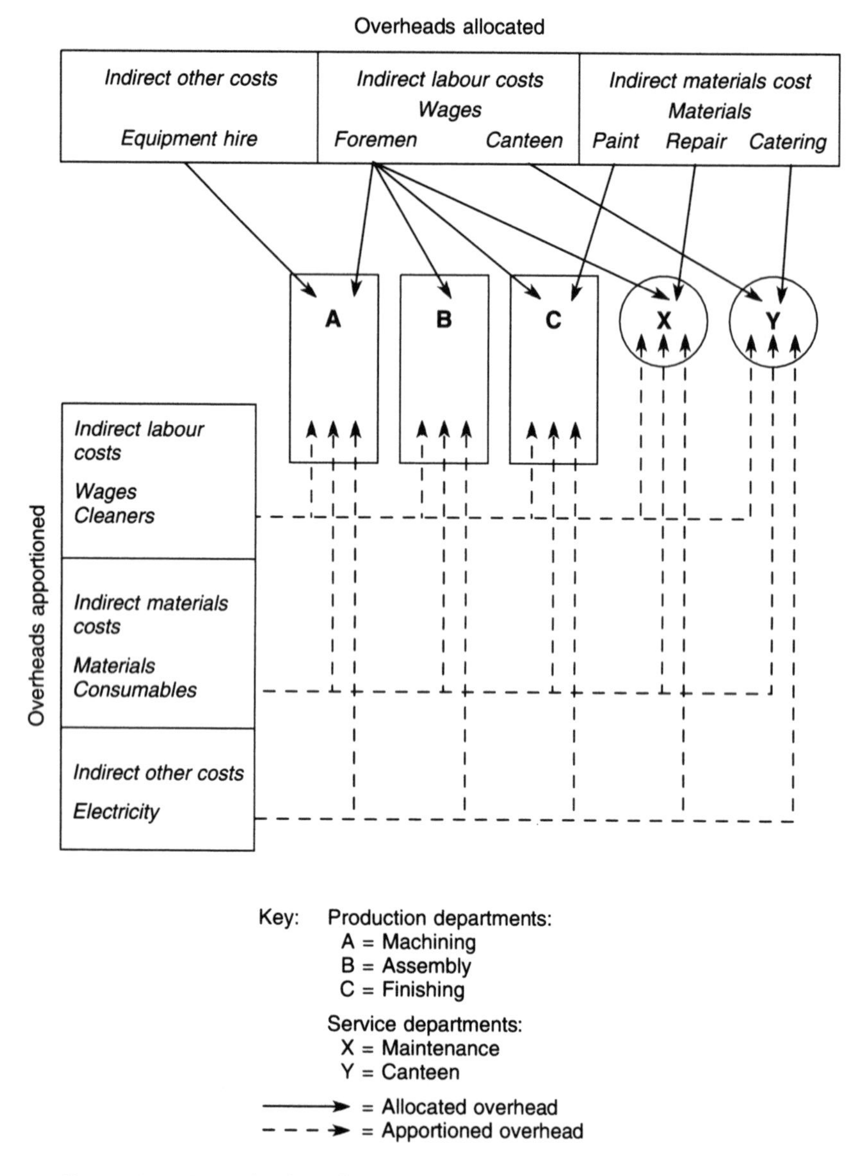

Figure 10.1 Example of production overhead allocation and apportionment

Table 10.1 Examples of bases of overhead cost apportionment

Overhead cost	*Usual basis of apportionment*
Rates	Area occupied
Heating	Volume of area occupied
Power	Horsepower/hours
Supervision	Number of personnel
Depreciation	Capital value
Sundry expenses	Predetermined percentage

Table 10.2 Example of the application of the principles of apportionment

			Department				
			A	*B*	*C*	*X*	*Y*
Basis		*Total*					
Area occupied (square metres)		5700	2 000	2 000	1 000	500	200
Volume occupied (cubic metres)		36 000	16 000	12 000	5 000	2 000	1 000
Horsepower/hours		310 000	250 000	50 000	5 000	5 000	–
Number of personnel		70	40	10	10	5	5
Capital value (£000)		410	305	45	32	20	8
Percentage (%)		100	41	27	22	7	3
Overhead cost	*Basis*	*£*	*£*	*£*	*£*	*£*	*£*
Rates	Area	45 000	15 789	15 789	7 895	3 948	1 579
Heating	Volume	11 000	4 889	3 667	1 528	611	305
Power	Hp/hrs	97 000	78 226	15 646	1 564	1 564	–
Supervision	Personnel	31 000	17 714	4 429	4 429	2 214	2 214
Depreciation	Capital value	53 000	39 427	5 817	4 137	2 585	1 034
Sundry expenses	Percentage	17 000	6 970	4 590	3 740	1 190	510

10.5 Stage II: Attribution of service department overheads to production departments

Ultimately, the production cost of a product must include all production-related overhead costs. For this reason, service department costs must be attributed to production departments. The manner in which this operation is carried out is by a similar process to that in Stage I as illustrated in Figure 10.1, that is, by a combination of allocation and apportionment.

Certain items can be allocated directly to production departments: for example, the cost of a replacement motor supplied by the maintenance department to a production department machine would be known and could be charged directly. But many other items of cost would not be so readily identifiable and would have to be apportioned on a suitable basis as illustrated in Tables 10.1 and 10.2 on p. 91. The Stage II process is represented diagrammatically in Figure 10.2.

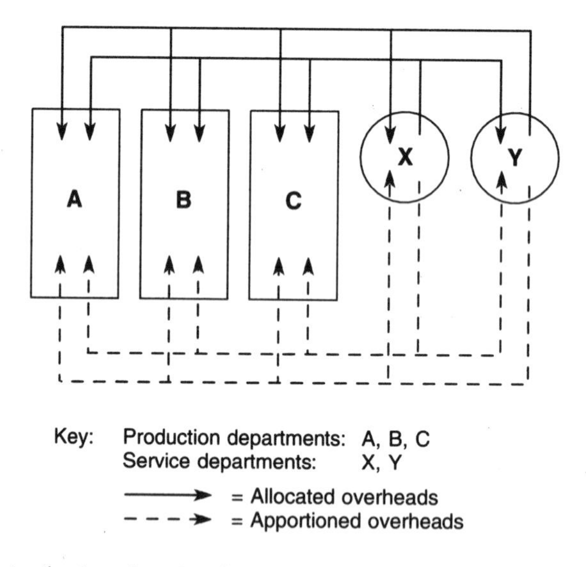

Figure 10.2 Attribution of service department overheads to production departments

One problem of the Stage II operation is that of cross charging of service department overheads. Scrutiny of Figure 10.2 reveals that Service Department X attributes its overheads to the three production departments and to Service Department Y, which itself attributes its overheads to the three production departments and to Service Department X. The charging and recharging of Service Department X by Service Department Y and vice versa, is, in theory, a never-ending process, but the purpose of Stage II is to transfer all the service departments' costs to production departments. In practice, it is relatively easy to achieve this objective arithmetically by a process of repeated distribution, provided that, as here, there are not many service departments. Where there are more service departments, an algebraic method has to be employed; where there are many, a computer programme has to be used, but these methods are outside the scope of this textbook.

10.6 Stage III: Attribution of production department overheads to production cost centres

At the conclusion of Stage II, all production overheads have been attributed to production departments, with none remaining in service departments.

It has to be realised that, within a single production department, various operations and/or processes are carried out. For example, in a production department in which four different operations are carried out, Product S may need to pass through three of them, while Product T may require only a single operation. In the interests of arriving at accurate product costs, products are charged with the overhead costs of only those operations and/or processes to which they are subjected.

The start point for achieving this objective is to regard each production department as a collection of mini departments, each performing different work. These are termed *cost centres*. Each cost centre is a group of homogeneous machines or processes. For example, within a particular production department there may be groups of drilling machines, boring machines and capstan lathes, each group of which would constitute a cost centre. Apart from machinery, a cost centre can also be area- or personnel-based, the overriding consideration being that it coincides with a sphere of responsibility, to enable effective control to be exercised. A cost centre consisting of, say, milling machines, would be the responsibility of a chargehand.

Taking into account the matters stated in the previous paragraph, it is possible to arrive at a more precise description:

- A *cost centre* is a location, a person, a machine or piece of equipment, or any group of these, for which costs are collected separately and used for cost ascertainment and control purposes.

At Stage III the overheads of each production department are attributed to the various cost centres within the department by means of the same system of allocation and apportionment as has already been seen at Stages I and II. This is illustrated in Figure 10.3 (on p. 94) in the form of a diagram. It is apparent from the diagram that, at the end of Stage III, all the production overheads which entered the system at Stage I have been attributed entirely to cost centres. There now remains only one further step before these overheads are attributed to products.

10.7 Stage IV: Absorption of cost centre overheads by individual products

The process by which cost centre overheads are attributed to products is termed *overhead absorption*; an alternative label is *overhead recovery*:

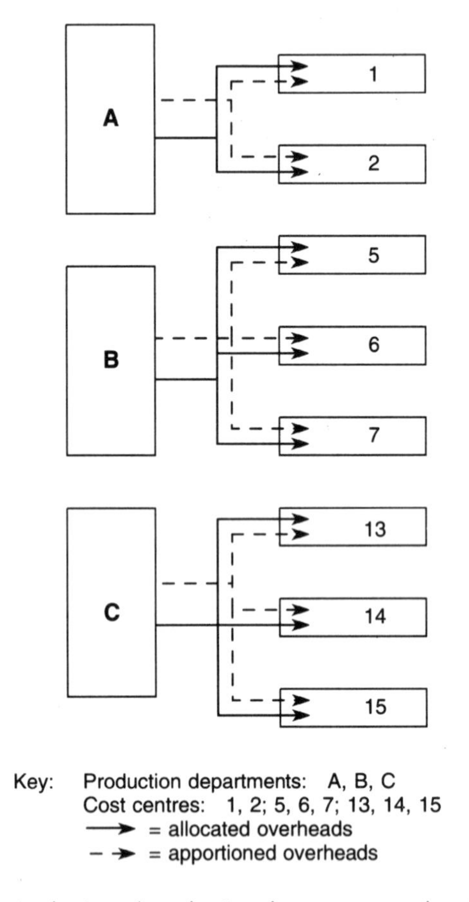

Figure 10.3 Attribution of production department overheads to cost centre

- *Overhead absorption* is the process by which individual products bear a proportion of the overheads of each cost centre through which they pass.

Different products pass through different combinations of cost centres prior to completion. This stage is illustrated diagrammatically in Figure 10.4. There are various methods by which the absorption process is carried out; these are the subject-matter of the remainder of this chapter.

It is apparent that products needed to pass through a different combination of operations or processes. Product P required work to be performed in cost centres 1, 5, 7, 14 and 15, while the manufacture of Product Q was carried out in cost centres 1, 2, 6, 7, 13 14 and 15. This fact will be reflected in the respective costs of the products.

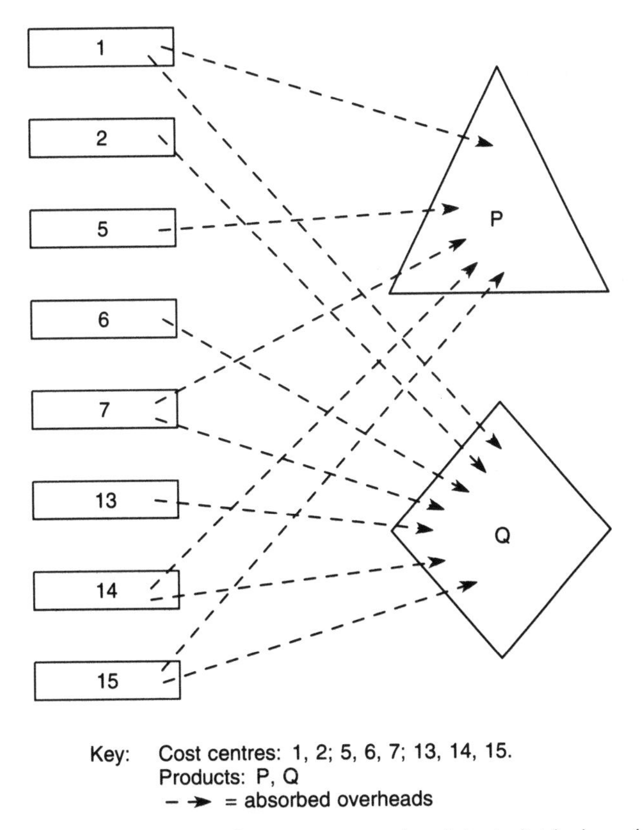

Figure 10.4 Absorption of cost centre overheads by individual products

At the end of Stage IV all the production overheads that entered the system at Stage I should, in theory, have been absorbed completely into product costs. In practice, however, this rarely, if ever, happens. There are various reasons why this is so. One reason is that, initially, some overhead cost figures have to be estimated in advance and the actual figures turn out to be different. Another reason is that the activity levels on which absorption rates (the subject of the remainder of this chapter) are calculated, do not coincide with the real levels of activity, thus leading to under- or over-absorbed overheads. Such differences are eventually accounted for within the profit and loss account of the business.

10.8 Overhead absorption methods

The final stage of the production overhead allocation process was described in Section 10.7, and illustrated in Figure 10.4 as being a matter of absorption.

There are various absorption methods available, the most common of which are considered in the paragraphs which follow. These are based on:

direct labour hours;
direct labour cost;
direct materials cost; and
machine hours.

The calculation of each of the above methods, their appropriateness under different circumstances and their application will each be examined.

10.9 Absorption – direct labour hour basis

This involves estimating the total number of labour hours available to each cost centre for a period (usually one year) and dividing the total estimated overhead cost of that centre for the same period by the number of hours. The resultant figure is an overhead absorption rate per direct labour hour which every product spends in the centre:

- $\frac{\text{Cost centre overhead costs}}{\text{cost centre direct labour hours}} = \text{direct labour hour overhead absorption rate}$

Example

Data for cost centre 7 estimated for year:

Cost centre overhead cost: £197 700
Number of direct labour hours: 5 250 hours

$$\frac{197\,700}{5\,250} = £37.65 \text{ per direct labour hour}$$

10.10 Absorption – direct labour cost percentage basis

The direct labour wage rates of the various grades of labour are applied to the totals of the corresponding estimated labour hours for the period in order to arrive at the total direct labour cost. The total estimated overhead cost of the centre for that period is expressed as a percentage of the direct labour cost. Overhead absorption is then a matter of applying the percentage to the direct labour cost of each product passing through the centre:

- $\frac{\text{cost centre overhead costs}}{\text{cost centre direct labour cost}} \times \frac{100}{1} = \text{direct labour cost percentage}$

Example

Data for cost centre 7 estimated for year:

Cost centre overhead cost: £197 700
Direct labour cost = £35 175

$$\frac{197\,700}{35\,175} \times \frac{100}{1} = 562\% \text{ of direct labour cost}$$

10.11 Absorption – direct materials cost percentage basis

The estimated overhead cost of the centre for the period is expressed as a percentage of the cost of the direct material cost content of all products passing through the centre. This percentage is then applied to the direct materials cost of the individual products in order to absorb the overhead cost.

- $\dfrac{\text{cost centre overhead costs}}{\text{cost centre direct materials cost}} \times \dfrac{100}{1} = \text{direct materials cost percentage}$

Example

Data for cost centre 7 estimated for year:

Cost centre overhead cost: £197 700
Direct materials cost: £217 600

$$\frac{197\,700}{217\,600} \times \frac{100}{1} = 90.9\% \text{ of direct materials cost}$$

10.12 Absorption – machine hour rate basis

The estimated overhead cost of the centre for the period is divided by the total number of machine hours estimated to be available for that centre. This calculation produces an overhead absorption rate which is applied to each machine hour which a product spends in the centre:

- $\dfrac{\text{cost centre overhead costs}}{\text{cost centre machine hours}} = \text{machine hour overhead absorption rate}$

Example

Data for cost centre 7 estimated for year:

Cost centre overhead cost: £197 700
Number of machine hours: 87 400 hours

$$\frac{197\,700}{87\,400} = £2.26 \text{ per machine hour}$$

10.13 Comparison of absorption methods

The four different absorption methods described in Sections 10.9 to 10.12 can lead to drastically different amounts of overheads being absorbed, as illustrated below.

Example

Data for Job Number 123 which passed through cost centre 7

Direct labour cost £74.20
hours 10 hours
Direct materials cost: £129.00
Machine hours: 165 hours

(The number of machine hours here (165), is considerably in excess of direct labour hours (10), indicating that each operative is responsible for running several machines simultaneously.)

Using the absorption rates already calculated in Sections 10.9 to 10.12, calculations can now be made using each basis in turn:

- *Direct labour hour basis*
  Cost centre 7 – Job No. 123 overhead absorption
  10 hours × £37.65 per hour = £376.50
- *Direct labour cost percentage basis*
  Cost centre 7 – Job No. 123 overhead absorption
  562.0% × £74.20 = £417.00
- *Direct materials cost percentage basis*
  Cost centre 7 – Job No. 123 overhead absorption
  90.9% × £129.00 = £117.26
- *Machine hour rate basis*
  Cost centre 7 – Job No. 123 overhead absorption
  165 hours × £2.26 per hour = £372.90

As is clearly shown by these examples, the amounts of cost centre overheads absorbed differ considerably in amount from £117.26 at one extreme to £417.00 at the other, with significant repercussions on total cost.

It follows, therefore, that a careful selection of overhead absorption methods must be made if accurate product costs are to be obtained. The procedure adopted is for the factors which give rise to the overhead costs to be examined: for example, their incidence may be governed by the passage of time, by direct cost, by level of activity, or by some other cause. In fact, the majority of overhead costs arise on a time basis; consequently one of the hourly rate methods is frequently the most appropriate. Almost invariably, this means the machine hour method, which is particularly appropriate where machine operations are automated. Direct labour hour or cost methods may, however, be suitable. This situation could arise in a cost centre where the use of machinery is minimal, the operation being carried out mainly by hand, by operatives using tools rather than machinery.

Care has to be exercised when using either of the direct labour methods. There is no difficulty when there is only one grade of labour within the cost centre, but problems arise when different grades of labour, earning different rates of remuneration, are used. The more highly skilled, and therefore more highly paid, operatives can presumably complete their tasks in less time than their lower grade co-employees. This has implications for the choice between a direct labour hour or a direct labour cost percentage method.

Although the calculation of the direct materials cost percentage basis has been illustrated, the number of instances where this is suitable are relatively rare.

11 Cost Ascertainment

11.1 Introduction

The purpose of cost classification and overhead attribution (described in Chapters 9 and 10) is to enable the cost of the individual items of output to be calculated.

Total costs of a business consist of prime costs – direct materials, direct labour and direct other costs – plus production overheads to give total production cost. To this figure is added the remaining overheads for selling, distribution and administration. The resultant grand total of cost can then be set against the sales value to disclose an overall profit or loss (see Table 11.1, which shows the estimated data of a business for a year). These figures provide the basis for absorption by individual products of the overheads of selling, distribution and administration.

Table 11.1 Build-up of total cost and profit (over a one-year period)

	£	£
Sales		987 000
less		
Direct labour	185 000	
Direct materials	41 000	
Prime cost	226 000	
Production overheads	437 000	
Production cost	663 000	
Selling and distribution overheads	119 000	
Administration overheads	98 000	
Total cost		880 000
Profit/(loss)		107 000

11.2 Absorption of non-production overheads

The analysis of overheads by function has been dealt with in Chapter 9. Examples of specific overhead costs to be classed as selling and distribution and as administration were given in Table 9.2. Absorption of production overheads was the subject-matter of Chapter 10, which illustrated the elaborate procedures that have to be adopted.

The methods by which non-production overheads are absorbed are fairly simple. It is usual for non-production overheads to be absorbed as a percentage of either sales value or production cost:

$$\frac{\text{Selling and distribution/administration overheads}}{\text{Sales value}} \times \frac{100}{1}$$

$$= \text{absorption percentage}$$

or

$$\frac{\text{Selling and distribution/administration overheads}}{\text{Production cost}} \times \frac{100}{1}$$

$$= \text{absorption percentage}$$

The relevant rates for the data given in Table 11.1 can be obtained by substituting the appropriate figures in the above formulae, as shown in Table 11.2.

Table 11.2 Calculation of non-production overhead absorption rates

Sales value basis

$$\frac{119\,000}{987\,000} \times \frac{100}{1} = 12.06\%$$ selling and distribution overhead absorption rate

$$\frac{98\,000}{987\,000} \times \frac{100}{1} = 9.93\%$$ administration overhead absorption rate

Production cost basis

$$\frac{119\,000}{663\,000} \times \frac{100}{1} = 17.95\%$$ selling and distribution overhead absorption rate

$$\frac{98\,000}{663\,000} \times \frac{100}{1} = 14.78\%$$ administration overhead absorption rate

Included in the data supplied in Table 11.1 is the data relating to Product G:

	£
Sales value	36 800
Production cost	14 100

Total cost and profit, using the figures calculated on each of the two bases for non-production overheads from Table 11.2, are shown in Table 11.3.

Table 11.3 Application of non-production overhead absorption rates

Sales value basis		
Product G	£	£
Sales value		36 800
less		
Production cost	14 100	
Selling and distribution overheads [12.06% × 36 800]	4 438	
Administration overhead [9.93% × 36 800]	3 654	
Total cost		22 192
Profit/(loss)		14 608
Production cost basis		
Product G	£	£
Sales value		36 800
less		
Production cost	14 100	
Selling and distribution overheads [17.95% × 14 100]	2 531	
Administration overhead [14.78% × 14 100]	2 084	
Total cost		18 715
Profit/(loss)		18 085

It can be seen from this illustration that the production cost basis has produced a greater amount of profit (£18 085) than the sales value basis (£14 608). This result must not be misinterpreted as indicating that the production cost absorption basis is preferable to its alternative. Product G is just one of the products of the business, and each has a different cost profile. The non-production cost overhead absorption percentages were calculated in Figure 11.2 on the aggregate data of the business; production cost (£663 000) constitutes 67.2 per cent of sales value (£987 000). Within those figures, however, the production cost of Product G (£14 100) is only 38.3 per cent of sales value (£36 800). From this fact it follows that the production cost (to which the absorption percentage is applied) of some of the other products, must be in excess of the overall 67.2 per cent of sales value. Thus, for some of the other products, production cost and total cost will be relatively higher than for Product G, and therefore profit will be lower. From a total business viewpoint, however, an identical amount of non-production overhead costs will be absorbed, irrespective of the basis selected for absorption, resulting in an identical amount of aggregate profit.

11.3 Cost units

Up to this point the narrative has been concerned with the ascertainment of the aggregate costs and profit of a business and of its homogenous products. Ultimately, however, costs are ascertained for *cost units*:

- A *cost unit* is a unit of production or service, expressed in quantitative terms, in relation to which costs are ascertained for planning and control purposes.

Businesses select cost units which are appropriate to the nature of the output or service of each part of the organisation (see Table 11.4).

Table 11.4 Examples of various outputs, services and their cost units

Type of output/service	*Suitable cost unit*
	(Single units)
Manufacture of special order	Per job
Manufacture of special orders of identical items	Per 1000 units manufactured or batch
Routine manufactures	Per 1000 units manufactured (litres – paint; tonnes – cement)
Catering	Per meal prepared
Professional services	Per chargeable hour (surveyor)[1]
Extractive industry	Per tonne or cubic metre extracted (sand quarry or gravel pit)
Plant hire/vehicle hire	Per day available[2]
	(Compound units)[3]
Electricity generation	Per kilowatt hour
Road haulage	Per tonne mile
Hospital	Per bed/day occupied
Road transport	Per passenger mile or kilometre

Notes

1. In a professional office, such as that of an architect, surveyor, accountant and solicitor, the total hours worked by all the staff greatly exceed the number of hours for which clients can be charged fees. This situation arises because the qualified, fee-earning staff need the back-up of support staff – clerks, typists, draftspersons, secretaries – and also because even the qualified staff are not engaged on fee earning work for the whole of their working hours. Under circumstances such as these, the cost of all hours worked is divided, not by total hours,

but by chargeable hours. This cost per chargeable hour can then be set against the fee per chargeable hour (that is, fees earned divided by chargeable hours) to disclose a meaningful figure of profit or loss per chargeable hour.

2. The start point for this is the number of days in the calendar year which coincides with the accounting year. From this figure is deducted days when plant is unavailable for hire: for example, weekends, statutory holidays and annual holidays (if the business closes down), together with days set aside for planned preventive routine maintenance. This leaves the number of available days which can be used as the cost unit for comparison with the hire charge per day, producing a profit or loss per available day.
3. Sometimes the use of a single cost unit – litres, tonnes, hours – is not really adequate for cost control purposes. This situation can arise where the sales revenue is derived from one set of circumstances but the costs are derived from a different set. It is then more appropriate to use a compound unit. A simple example should make this clear. If a bus company operates a scheduled route, the sales revenue consists of the fares paid by passengers, whereas the operating cost is governed by the distance travelled and is incurred whether any fare-paying passengers board the bus or not. Under these circumstances, it is more meaningful to relate both the passenger fares and the distance generated costs to the same unit, termed, in this case, passenger kilometres, to reflect the effect on profit or loss of the number of passengers carried.

The data relating to a single journey undertaken by a vehicle operated by a road haulage company is given in Table 11.5.

Table 11.5 Vehicle operating data

Distance travelled	
Depot to destination	25 kilometres
Destination back to depot	25 kilometres
Total	50 kilometres
Haulage charges	£8 per tonne
Haulage costs	
Drivers' wages	£22 for journey
Diesel fuel used	£13 for journey
Garage overheads	£0.60 per kilometre

Calculate the operating profit or loss on each of the following two bases:

1. The vehicle carried a load of 4 tonnes from the depot to the destination and returned empty to the depot.
2. As for 1., except that on the return journey to the depot it brought back a load weighing 6 tonnes. Ignore the fact that the extra weight might result in a slightly increased diesel fuel usage.

$$\text{tonne/kilometres} = \frac{\text{tonnes carried} \times \text{kilometres covered}}{2}$$

The comparative effects of the use of a single cost unit and a compound cost unit are illustrated in Table 11.6.

Table 11.6 Comparison of the effect of a single and a compound cost unit

	Basis 1			*Basis 2*		
Distance covered	*50 km*			*50 km*		
Weight carried	*4 tonnes*			*10 tonnes*		
Tonne/kilometre	*100 tonne/km* [(4 × 50)/2]			*250 tonne/km* [(10 × 50)/2]		
	Amount £	*Per km* £	*Per tonne/km* £	*Amount* £	*Per km* £	*Per tonne/km* £
Earnings [4 × £8; 10 × £8]	32	0.64	0.32	80	1.60	0.32
less						
Operating costs						
Wages	22	0.44	0.22	22	0.44	0.09
Diesel oil	13	0.26	0.13	13	0.26	0.05
Garage overheads	30	0.60	0.30	30	0.60	0.12
Total costs	65	1.30	0.65	65	1.30	0.26
Operating profit				15	0.30	0.06
and (loss)	(33)	(0.66)	(0.33)			

From Table 11.6 it is apparent that the cost per km (£1.30) is the same, whether the vehicle returns empty to the depot or not. In other words, the simple cost unit completely ignores differences in loading. The tonne/km cost unit overcomes this deficiency by spreading the operating costs, which, apart from the fixed element of the garage overheads, would not have been incurred if the vehicle had not made the journey, over the load carried. The greater the load carried, the less the tonne/km unit cost.

It is apparent from Basis 1 that when the vehicle is carrying a load in one direction only, the tonne/km unit cost is £0.65, resulting in a loss of £0.33 per tonne/km, whereas when loaded for the return journey as well, operating cost is reduced to £0.26 per tonne/km and a profit of £0.06 per tonne/km arises. This is a far more accurate indication of the commercial reality of the situation than the use of a single 'per km' cost unit would provide.

11.4 Job cost ascertainment

The practices and procedures contained in Chapter 10 and in the earlier parts of this chapter can now be employed in the ascertainment of job costs, an example of which is given in Table 11.7. The data is available for Product F for Period 7/12.

Table 11.7 Example of a job cost/profit statement for Product F for Period 7/12

Number of units produced	10 367 units
Machine hours utilised	17 461 hours
	£
Selling price per unit	24
Direct materials cost	80 204
Direct labour cost	12 098
Direct expenses	1 145
Overhead absorption bases:	
Production	£6.31 per machine hour
Selling and distribution	6.34% of production cost
Administration	4.15% of production cost

Cost statement

Output: 10 367 units

	Amount £	*Per unit* £
Sales value	248 808	24.00
Costs		
Direct materials	80 204	7.73
Direct labour	12 098	1.17
Direct expenses	1 145	0.11
Prime cost	93 447	9.01
Production overheads [17 461 machine hours @ £6.31 per hour]	110 179	10.63
Production cost	203 626	19.64
Selling and distribution overheads [6.34% × 203 626]	12 910	1.24
Administration overheads [4.15% × 203 626]	8 450	0.82
	21 360	2.06
Total cost	224 986	21.70
Profit/(loss)	23 822	2.30

11.5 Process costing

Particular difficulties arise in ascertaining unit costs where processes are involved, as in the manufacture of paint, for the following reasons:

1. By nature they constitute a series of continuous processes,
2. Unlike ordinary production jobs, process units are part of a homogeneous mass in which individual units are indistinguishable.
3. Units of raw material which form the input of the first process are frequently supplemented by further units of the same and/or different units in subsequent processes.
4. In some or all processes losses can occur for various reasons, including evaporation and chemical reaction. Such losses may be:
   (a) normal, in which case they are regarded as part of the cost of the process concerned; or
   (b) abnormal, and written off against the operating account.
   For similar reasons, normal and abnormal gains can also arise and receive the same accounting treatment.
5. At the end of an accounting period completed units will usually have been transferred to the next process or to completed stock, if at the end of the last process. By the nature of the continuous production processes, certain units will be incomplete, in varying degrees. Process costs, however, are incurred on all units, both complete and incomplete. For cost purposes, a figure of equivalent units has to be calculated to enable the unit cost of the process to be calculated.
6. Processing frequently gives rise to joint products and by-products. Joint products arise simultaneously from the same process but, up to the point of their separation are inseparable. They often require further processing from that point and are usually valuable in relation to the other process products. Examples include coke, gas and naphtha in coal processing, and petrol, diesel and lubricants in oil refining. By-products are also produced during the main processes but are incidental to the main products and are relatively less valuable. Sawdust is a by-product of the timber trade and furnace slag from steelworks is used in highway foundations and cement manufacture.

The data shown in Table 11.8 relates to the production of a synthetic chemical in Process 2 in Period 8/12. All quantities are in units of 1000 litres. The number of units of input into Process 2 in Period 8/12 and completed in period 8/12 is 460. It follows that the total costs above relate to the 460 units started and completed during the period, plus, not only the processing work needed to complete the opening work in process, but also that carried out on the closing figure of work in process.

Calculations must be made to arrive at an equivalent units figure. Taking labour as an example, opening work in process relates to 10 units which are 60 per cent complete; this is regarded as equal to 6 completed units. A further 40 per cent of total labour is required before the units are complete, which is equivalent to the labour cost of 4 completed units. The same reasoning is applied to the other items of cost. The calculation of Process 2 cost in Period 8/12 is shown in Table 11.9 and of closing work in process and completed units in Table 11.10.

Table 11.8 Calculation of process unit costs

Work-in-process

	Period 8 opening		*Period 8 closing*	
	Number of units	*Percentage completion*	*Number of units*	*Percentage completion*
Labour	10	60	20	40
Materials	10	70	20	60
Process overheads	10	60	20	40

Total cost for Period 8/12 for Process 2

	£
Labour	6 300
Materials	19 500
Process overheads	10 200
Total for Process 2	36 000

Table 11.9 Calculation of Process 2 cost in Period 8/12

Cost item	*Started and completed in Period 8/12*	*Work in process equivalent units completed in Period 8/12*		*Total units*	*Total cost*	*Cost per 1000 litres*
		Opening	*Closing*			
	1000 litres	*1000 litres*	*1000 litres*	*1000 litres*	*£*	*£*
Labour	460	4 [40% × 10]	8 [40% × 20]	472	6 300	13.30
Materials	460	3 [30% × 10]	12 [60% × 20]	475	19 500	41.10
Process overheads	460	4 [40% × 10]	8 [40% × 20]	472	10 200	21.60
Totals					£36 000	£76.00

Table 11.10 Calculation of Process 2 closing work in process and completed units transferred in Period 8/12

Equivalent units of closing work in process:

	Number of units	*Cost per unit*	*Value (to nearest £1)*
	1000 units	£	£
Labour	8	13.30	106
Materials	12	41.10	493
Process overheads	8	21.60	173
			772

	£
Total cost in Period 8/12	36 000
less	
Value of closing work in process	772
Cost of completed units transferred to Process 3 in Period 8/12	35 228

11.6 Valuation of stocks

The financial accounting treatment of stock has already been dealt with in Chapter 4, where is was seen that the types of stock held vary with the nature of the business concerned.

A business in the building and construction industry will hold very diverse items of stock. A typical example can be seen in Appendix I, where the annual accounts of John Maunders Group plc disclose four different types of stock:

- Land for development;
- Building materials;
- Work-in-progress; and
- Houses for resale.

As was noted in Chapter 4, the valuation placed on stocks, other than long term contract work in progress (which is the subject of Section 11.7), is governed by the combined requirements of Schedule 4 of the Companies Act 1985 and by the accounting standard SSAP 9:

- *Stocks* should be stated in financial statements at the aggregate of the *lower of cost and net realisable value* of the separate items of stock or of groups of similar items.

The terms *net realisable value* and *cost* need to be fully understood.

- *Net realisable value* means:
  Selling price (actual or estimated if actual is not known)
  less (as applicable)
  trade discounts;
  all further costs needed to make item complete and saleable; and
  all costs to be incurred in marketing, selling and distribution.

- *Cost* is defined as expenditure incurred in the normal course of business in bringing the stock to its present condition and location. Expenditure therefore includes:
  purchase cost; and
  appropriate conversion costs.

These terms are further explained:

- *Purchase cost* comprises:
  basic list price cost
  plus (as applicable)
  import duties;
  handling costs;
  transport costs; and
  other directly attributable costs
  less (as applicable)
  trade discounts;
  rebates; and
  subsidies.

- *Conversion costs* are stated to include:
  direct labour;
  direct materials;
  direct expenses;
  sub-contracted work;
  plus (as applicable);
  production overheads; and
  other relevant attributable overheads.

- *Production overheads* means those overheads classified under the production function, as outlined in Chapter 9, in Tables 9.1 and 9.2, based on the normal annual level of activity and includes specifically depreciation of assets used in production.

In addition to the above, it is sometimes permissible to include loan interest as a component of stock cost where it can be clearly shown that the loan, on which the interest has been incurred, has been borrowed specifically in connection with the stock item(s) concerned.

11.7 Valuation of long-term contract balances within stocks

Chapter 7 dealt in detail with the compilation and disclosure requirements of the accounting standard SSAP 9 with regards to the profit and loss account and balance sheet.

In the balance sheet, within the heading 'Stocks', the item 'long-term contract balances' is, in effect, work-in-progress, and is included on the basis specified by SSAP 9.

- Long-term contract balances – costs incurred to date minus the aggregate of:
  amounts transferred to cost of sales in the profit and loss account;
  foreseeable losses; and
  payments on account not matched with turnover.

In this context, costs incurred are as defined in Section 11.6 above, the other terms having been explained in Chapter 7.

11.8 Absorption cost basis

Analysis into functions and according to behaviour was the subject matter of Chapter 9, where it was seen that costs can be fixed, variable or semi-variable. This theme was developed in Chapter 10, which set out the various stages needed before production overheads could be absorbed by units of output. Chapter 11 has progressed further to show in Table 11.7 the cost statement for a product.

So far in this book, and in dealing in this chapter, in Sections 11.6 and 11.7, with the valuation of stocks and with long-term contract balances, the basis on which the figures have been prepared has been that both the fixed and variable elements of the costs (in other words, *all* the costs) have been absorbed. This, for obvious reasons, is termed *absorption costing* and is the normal practice for financial accounting purposes for external disclosure and presentation of figures.

11.9 Marginal cost basis

For internal reporting, planning and decision-making purposes, a different basis, known as *marginal costing* is employed. Marginal costing deals only with variable costs, as these vary with levels of output, whereas fixed costs are more governed by time – rent, rates, insurance, for example – than by level of activity; thus, being a time cost they are incurred whether productive activity is taking place or not and, being unaffected by changes in activity levels, are irrelevant in many policy decision situations. The use of marginal costing in short-term decision-making situations is dealt with in Chapter 13.

12 Cost/Volume/Profit Analysis

12.1 Introduction

As has been seen in earlier chapters, the profit or loss made by a business is the difference between its sales revenue and its costs, both fixed and variable. The amount of profit earned or loss sustained varies according to the level of activity. An examination of this relationship between profit or loss, sales and costs is important to the managers of a business in assessing the effects of their current, planned or proposed policies. This relationship can be explored for different levels of activity either by calculation or by graphic means, both of which are explained in the remainder of this chapter, using cost/volume/profit (CVP) analysis.

12.2 Break-even analysis

The graphic version of CVP analysis is to be found in what is termed break-even analysis, so called from the chart produced to show the relationships.

Three of the components of a break-even chart have already been encountered in Chapter 9, where there was depiction in graph form of fixed overheads in Figure 9.1, of variable overheads in Figure 9.2, and of semi-variable overheads in Figure 9.3. The fourth item of a break-even chart is the sales line which, like the variable overheads line, starts from zero and ascends in a straight line. These four items are combined into a single graph, in one of two versions. A common method is for the fixed cost line to be drawn first and the variable costs, comprising direct costs, variable overheads and the variable element of semi- variable overheads, to be drawn at the point where the fixed cost line cuts the vertical axis. This is illustrated in Figure 12.1. The alternative method is for the variable cost line to start at zero and the fixed cost to be drawn as a line parallel to the variable cost line. This is illustrated in Figure 12.2. In either case, the area within the two cost lines in combination constitutes total cost. The point at which the sales revenue line intersects the total cost line is termed the break-even point.

- The *break-even point,* which can be measured on the horizontal axis in terms of activity and on the vertical axis in value terms, shows the point at which the business makes neither a profit nor a loss. Up to that point the business makes a loss; and beyond it a profit.

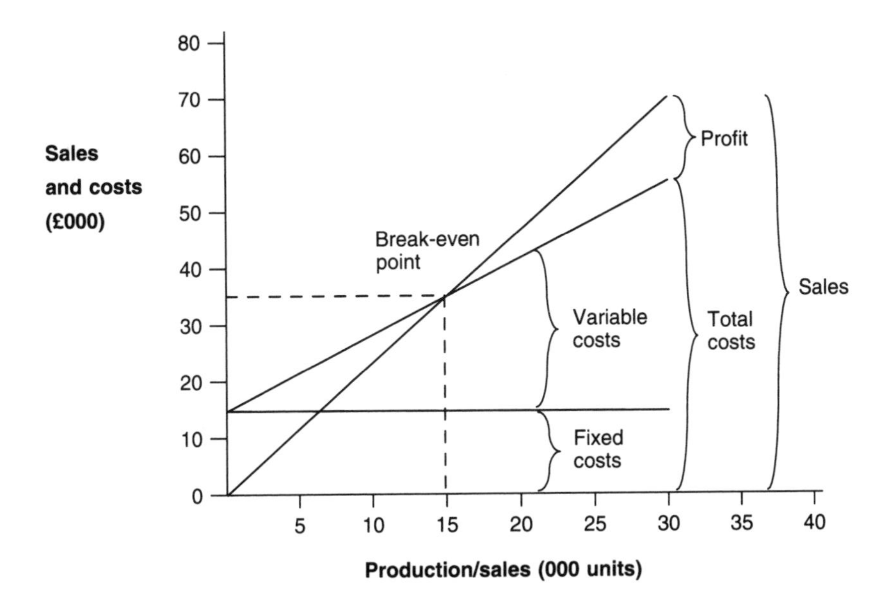

Figure 12.1 Break-even chart – first version

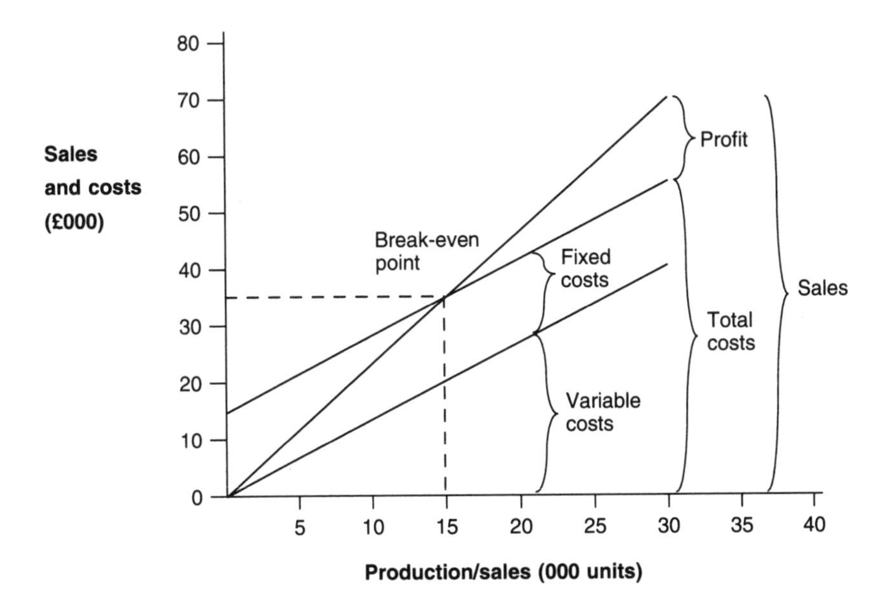

Figure 12.2 Break-even chart – second version

It can be seen from Figure 12.1 that the break-even point occurs when production/sales reaches 15 000 units with a sales value, or at a cost, of £35 000.

Break-even point in Figure 12.2 is in a similar position to that in Figure 12.1. In both graphs it is apparent that, with an output of 30 000 units being produced and/or sold, the business is operating at below its full capacity of 40 000 units.

12.3 Break-even charts – other aspects

Other useful pieces of information can be obtained from a break-even chart, in addition to the position of the break-even point itself. These include the *angle of incidence* and the *margin of safety*.

- The *angle of incidence* is the angle formed by the intersection of the sales and total costs lines. The width of the angle indicates the rate at which the business makes a profit once break-even point has been exceeded. Conversely, it shows the rate at which losses are incurred below the break-even point. In both cases, the wider the angle, the faster the rate at which profits are earned, or losses sustained, as the case may be.
- The *margin of safety* is the difference between the break-even units and the output being produced or sold, expressed either as a number of units or as a percentage of total units. It indicates the degree to which a business is vulnerable to changes in levels of activity. A narrow margin shows that a business is unable to absorb a significant reduction in the level of activity and still remain in profit. Businesses which have a relatively high level of fixed costs are usually found to have a narrow margin of safety. This situation arises in those businesses which have a heavy investment in fixed assets with a correspondingly high level of fixed costs. There are various examples of this situation in the iron and steel manufacturing sector and in highly automated businesses.

The angle of incidence and the margin of safety are illustrated on the break-even chart shown in Figure 12.3

From the graph in Figure 12.3 it is apparent that the maximum capacity is 40 000 units, but that the business is operating at 75 per cent of this figure, at 30 000 units of production/sales. Break-even point occurs at 15 000 units, leaving a margin of safety of 15 000 (30 000 – 15 000 break-even units) between the break-even point and the activity level. This can also be expressed as a margin of safety of 50 per cent $[(30\,000 - 15\,000)/30\,000 \times 100/1]$.

A break-even chart enables a user to obtain a quick visual appreciation of the cost/volume/profit relationships within a business. It is then very easy to see immediately the effects of any changes in these relationships. For example, suppose the effect of a price increase needed to be assessed. An

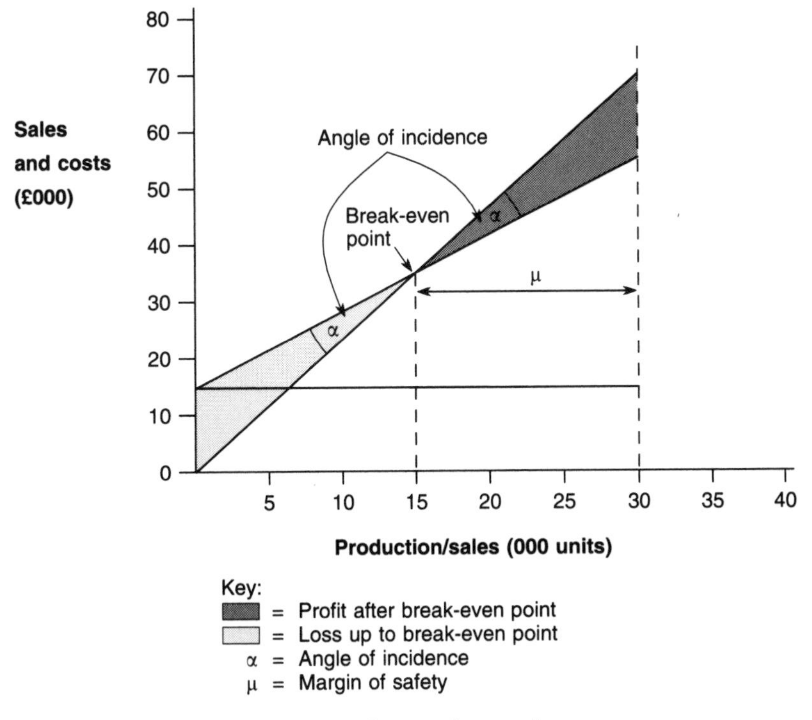

Figure 12.3 Break-even chart – other aspects

extra line could be inserted for the new amount of sales revenue; its effect on the break-even point and on the margin of safety would be apparent immediately. Similarly, changes in fixed and/or variable costs can be superimposed on the chart and their effects seen readily.

12.4 Profit/volume charts

A modified version of the break-even chart is one which illustrates the CVP relationship from a different perspective and is known as a profit/volume (P/V) chart. The horizontal axis represents sales in either unit quantity or in value. The vertical axis is intercepted by the horizontal axis at break-even, that is, at zero profit/loss. Above the horizontal axis is profit, below it is loss. The profit volume line starts below the horizontal axis at the point on the vertical axis equal to the fixed costs, indicating that when no production is taking place the loss is equal to fixed costs incurred. This line then moves in an oblique manner to intersect the horizontal axis at break-even point and continues upwards to the extent of the profit made. The break-even chart in Figure 12.3 could be redrawn as a profit/volume chart, as shown in Figure 12.4.

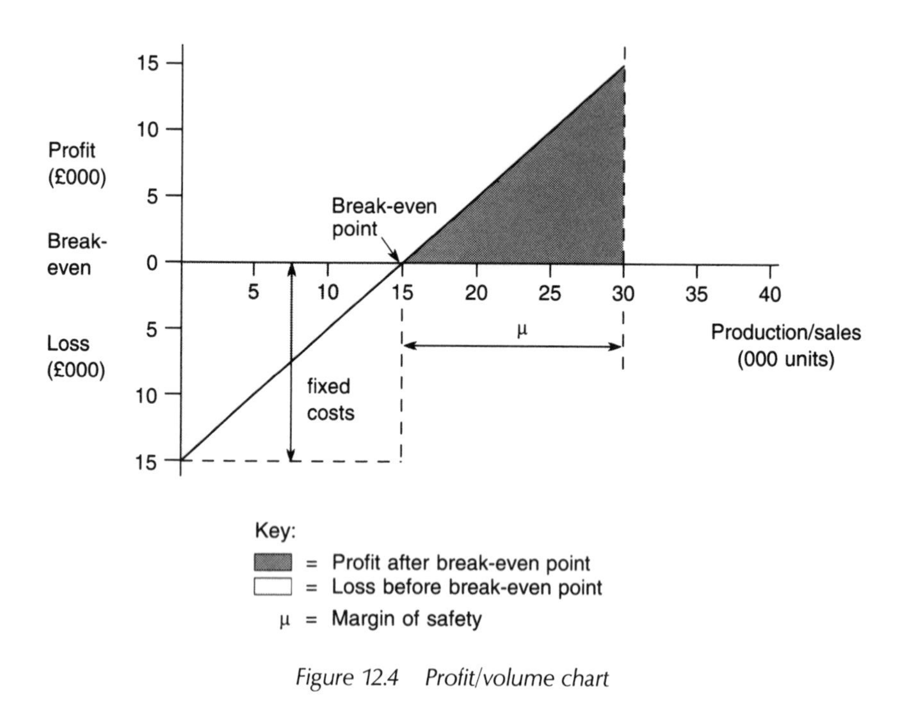

Figure 12.4 Profit/volume chart

12.5 Break-even and profit/volume charts – limitations

While these charts are, as has been seen, useful tools that enable management to gain an appreciation of the CVP relationship at a glance, they suffer from a number of limitations, of which users must be aware in order to avoid placing an over-reliance on the results:

(i) When used predictively, as in the previously noted consideration of policy changes options, the results suffer from the uncertainties connected with all forecast figures.

(ii) It can apply only to a single-product business, or one where the product mix is fixed: that is, the various products remain in fixed proportions. In other, multi-product, businesses, the results indicated by a break-even chart would be invalid because it would not be feasible to produce a separate chart for each product, because the attribution of fixed costs to products would have to be on a very arbitrary basis; a single chart covering all products would be invalid because of the changing mix of products, each with different selling prices and variable costs per unit.

(iii) It is assumed that variable costs and selling prices, in each case per unit, remain constant at all levels of activity; as a result they are drawn on the charts as straight lines. Unit costs can often be reduced, as higher activity levels are achieved, because of economies of scale. Similarly, selling prices per unit are usually reduced for bulk orders. Both of these circumstances would need to be depicted as curvilinear, but as the charts are regarded as a rule of thumb guide, this fact is almost invariably ignored.

(iv) Production and sales are assumed to be at the same activity level. Where this is not the case and part of the production is transferred to stock or, conversely, sales exceed production, leading to a stock reduction, the results indicated by the chart are unreliable.

12.6 CVP analysis by calculation

An alternative to the construction of a break-even chart is for the CVP relationship to be ascertained by various calculations centring on what is termed the contribution.

- *Contribution* is the figure of sales minus all variable costs – direct costs, variable overheads and the variable portion of semi-variable overheads. It can be expressed as an absolute figure, or per unit.

Using facts elicited from the break-even chart in Figure 12.1, at the activity level achieved (30 000 units) contribution is:

	Amount £	*Per unit* £
Sales	70 000	$2.\dot{3}$
less		
Variable costs (£55 000 total costs less £15 000 fixed costs)	40 000	$1.\dot{3}$
Contribution	30 000	1.0

Now follow formulae for the various CVP calculations, accompanied by calculation of the figures concerned, using the data obtained from Figure 12.1.

- *Break-even point in quantity terms*

$$\frac{\text{Fixed costs in total}}{\text{Contribution per unit}} = \text{Break-even point units}$$

$$= \frac{£15\,000}{£1} = 15\,000 \text{ units}$$

- *Break-even point in value terms*
  As a preliminary step, the contribution/sales ratio (C/S ratio) must be established:

$$\frac{\text{Contribution per unit}}{\text{Selling price per unit}} = \text{C/S ratio}$$

$$= \frac{£1.00}{£2.3\dot{3}} = 0.42857$$

$$\frac{\text{Fixed costs in total}}{\text{C/S ratio}} = \text{Break-even point value}$$

$$= \frac{£15\,000}{0.42857} = £35\,000$$

These calculations can be adapted to supply other information. For example, if the business needed to find out what level of production/sales would be required to produce a given amount of profit, the answer can be obtained by a slight modification of the break-even formula, thus:

$$\frac{\text{Fixed cost in total plus required profit}}{\text{Contribution per unit}}$$

$$= \text{level of activity to produce required profit}$$

Assuming the required profit is £20 000, the calculation becomes

$$\frac{£15\,000 + £20000}{£1.00} = 35\,000 \text{ units}$$

This indicates that the proposal is feasible, because the production/sales needed to produce a profit of £20 000 is within the maximum capacity of 40 000 units.

The resultant figure of 35 000 units can be proved to be correct by means of an independent calculation:

Production/sales =	*35 000 units*
	Amount £
Sales (35 000 × £2.3̇)	81 666
Variable costs (35 000 × £1.3̇)	(46 666)
Fixed costs	(15 000)
Required profit	20 000

13 Short-term Decision-making Techniques

13.1 Introduction

Throughout the life of a business, the management is regularly confronted by situations on which decisions must be reached. At lower management levels, among foremen and chargehands for example, the decisions have a limited, localised effect. Decisions which have wide-ranging implications for a business, its customers and clients are taken by the higher management levels of executives and directors.

Some of the decisions taken are regarded as being short-term, by which is meant that they are only intended to operate for a short period of time, or that they are such that they can be altered quickly if the desired objectives are not being achieved. This chapter is concerned with decision situations which fall into this category. Decision-making where the effects are long term are the subject matter of Chapter 14.

13.2 Relevant costs and revenues

Short-term decisions of the type described above are arrived at on the basis of relevant costs and revenues. 'Relevant' in this context means that the items concerned change according to the alternative courses of action. Decisions are based on the direction and amount of change. Certain items are unaffected by a particular decision and are therefore not relevant. Examples of these are *costs already incurred* (often termed *sunk costs*), *notional and other non-cash costs*, including *notional interest* and *rent and depreciation charges*.

From this, therefore, it can be seen that relevant items can be identified as such if they affect future cash flows.

13.3 Differential and incremental costs and revenues

As has been stated in the previous paragraph, in reaching decisions those costs and revenues which will have an impact on future cash flows are regarded as being relevant. Those costs and revenues which give rise to increased or reduced future cash flows are termed *differential*; *incremental* is the term used in the case of increased costs and revenues.

13.4 Opportunity cost

Where there are two or more alternative courses of action, opportunity cost is the cost of rejecting or forgoing the most favourable alternative. A simple example can illustrate this concept. Suppose that a business has a machine which can be used to manufacture units producing a positive contribution of £1700, or it could be hired to another manufacturer at a rental of £2 100, resulting in both cases in a nil residual value; alternatively, the machine could be sold now for its scrap value, £500. If the company should decide to scrap the machine, the opportunity cost of this decision would be £2 100, the value of the higher of the two non-selected alternatives. Applying this same principle, the opportunity cost of renting the machine to another manufacturer would be £1700, the loss of the positive contribution from use. Opportunity costs, together with differential costs, can be regarded as relevant costs.

13.5 Common short-term decision situations

There are a number of situations commonly encountered to which short-term decision-making techniques are applied to arrive at an optimum solution. These include the following situations:

a) product pricing;
b) special order acceptance;
c) segmental closure;
d) hire or buy;
e) make or buy; and
f) limiting factor constraints.

Each of these will now be explained and illustrated.

13.6 Product pricing

One problem facing a manufacturing or wholesaling company is the price at which it should sell its goods. Apart from the overriding need not to sell at a loss, except in special circumstances, the factors which affect prices are supply and demand, and competitors' prices. Companies evaluate different pricing strategies in order to arrive at an optimum: that is, the one which will produce the highest profit, as shown in the following example:

Example

A company made the following forecast of results for the following year:

Forecast
Sales 100 000 units

	Per unit £	*Amount* £
Sales	12.00	1 200 000
less Variable costs		
Materials	4.30	430 000
Labour	3.20	320 000
Overheads	1.70	170 000
Total	9.20	920 000
Contribution	2.80	280 000
less		
Total fixed costs		95 000
Net profit/(loss)		185 000

In considering this forecast, the directors are examining the effects of different prices. Market research has shown that demand for this product is price-sensitive. The final decision is to be based on the price which produces the highest figure of profit from the following three alternatives:

Alternative 1 To reduce the price to £10.50 per unit. This would give the company a competitive advantage and would enable it to operate at its maximum capacity of 130 000 units, all of which could be sold. Extra staff would have to be engaged, to cope with the increased demand, at a cost of £14 000, but variable costs per unit would not be affected.

Alternative 2 To increase the price to £13.00 per unit. This would meet with some consumer resistance, resulting in a reduction in sales to 87 000 units. Neither variable costs per unit nor total fixed costs would be affected by this decision.

Alternative 3 To adopt the forecast without alteration.

Solution

The short-cut method of calculating the effect on net profit is to calculate the effect on contribution and, if relevant, on fixed costs. The resulting figure gives the increase or decrease in net profit, thus:

Alternative 1, to reduce the unit price to £10.50:

	£	£
Forecast net profit		185 000
Alternative: contribution [130 000 × (10.50 – 9.20)]	169 000	
: extra fixed costs	(14 000)	
	155 000	
less Forecast contribution	280 000	
Reduction in forecast net profit		(125 000)
Alternative net profit		60 000

Alternative 2, to increase the unit price to £13.00:

	£	£
Forecast net profit		185 000
Alternative: contribution [87 000 × (13.00 – 9.20)]	330 600	
less Forecast contribution	280 000	
Increase in forecast net profit		50 600
Alternative net profit		235 600

If the company were to adopt the forecast unaltered, a net profit of £185 000 would result. However, an examination of the evaluation of the two alternatives clearly reveals that the price increase, Alternative 2, will maximise profit.

13.7 Special order acceptance

Occasionally businesses receive enquiries that would lead to the placement of an order on non-standard terms. This usually means that the potential customer is only offering to pay a price well below what is normal. The initial reaction might be to reject the enquiry outright but there are circumstances where it might be financially beneficial to convert it into an order.

If the price being offered exceeds the marginal cost, that is, if it would produce a positive contribution, it would be advantageous for the business to accept the order on this basis, assuming that fixed costs are absorbed by normal activity output. This is the basis on which bus and train excursion fares are calculated. All the fixed costs are absorbed by the normal scheduled services. Provided the excursion fares exceed the marginal costs of operating them, there will be a positive contribution which will increase overall net profit.

Example

The company illustrated in Section 13.6 received an enquiry from an overseas customer for the supply of 25 000 units. The customer was prepared to pay only £10.20 per unit. The company has sufficient capacity to accept the order. Fixed costs would not be affected but the forecast variable costs would be increased by the following amounts:

	Per unit £
Additional	
Packaging	0.30
Overtime	0.25
Transport	0.15
Total	0.70

Solution

It is a fairly simple matter to calculate the financial effect of accepting the order:

	Per unit £	*Amount* £
Sales (25 000 units)	10.20	255 000
less		
Marginal costs (9.20 + 0.70)	9.90	247 500
Contribution	0.30	7 500

On this basis, acceptance of the order would increase overall profit by £7 500. Even if extra fixed costs had been involved, it would still have been profitable provided they did not exceed the amount of the marginal contribution.

At this point it is appropriate to note that companies sometimes accept special orders even if a loss results. Reasons for doing so include the desire to gain a foothold in a particular market and/or the expectancy that this will lead to regular future orders.

There are, however, dangers involved in differential pricing. For some years the UK car manufacturing industry has been the target of criticism for its pricing policies, whereby a particular make and model of a vehicle can be bought much cheaper on the Continent than in the UK. Continental buyers are in effect being subsidised by the UK buyers, to the intense annoyance of the latter.

13.8 Segmental closure

In all businesses except the smallest, the various activities are organised into segments coinciding with spheres of management responsibility. These segments can take a variety of forms. In a merchandising concern the individual branch shops can be regarded as separate segments, as can individual departments within a large store. The segments of a manufacturing company are often designated as *divisions*. These divisions may be on the basis of class of business and/or geographically according to sphere of operations. Redland plc, for example, identifies its main operating divisions as Roofing, Aggregates and Bricks, and Other (which includes Plasterboard).

Whatever the size of the segment, it is usual for their sales, costs and profit or loss to be reported separately. Policy decisions are made on the basis of this, including decisions regarding their continuance or closure.

Example

The summarised operating statement of a business for the previous year disclosed:

Operating statement for the year ended 31 December Year X4

			Departments		
	A	*B*	*C*	*D*	*Total*
	£000	*£000*	*£000*	*£000*	*£000*
Sales	272	144	313	526	1 255
less					

Table continued next page

Table continued

Costs					
Variable	190	85	206	418	899
Fixed	51	27	46	159	283
Total	241	112	252	577	1182
Net profit/(loss)	31	32	61	(51)	73

The directors are proposing to close down Department D on the grounds that by doing so the company's total profit would have increased to £124 000 by the avoidance of the £51 000 loss, but require confirmation before the final decision is made.

Solution

Problems of this nature are solved by redrafting the figures in marginal costing format:

Operating statement for the year ended 31 December Year X4

			Departments		
	A	*B*	*C*	*D*	*Total*
	£000	*£000*	*£000*	*£000*	*£000*
Sales	272	144	313	526	1 255
less					
Variable costs	190	85	206	418	899
Contribution	82	59	107	108	356
less					
Fixed costs					283
Net profit/(loss)					73

Examination of the above statement reveals that all departments have made a positive contribution and that, at £108 000, the contribution of Department D, whose closure is under consideration, is the highest of all four departments. Closure of Department D would have resulted in the loss to the business of this contribution of £108 000. Furthermore, there would not have been a reduction in total fixed costs; the £159 000 fixed costs allocated to Department D in the example would have had to be borne by the remaining Departments A, B and C.

The implication of the above matters for the company as a whole would be that the net profit of £73 000 would have become a net loss of £35 000 by reason of the loss of Department D's contribution, £108 000. Viewed from the perspective of the original example, if Department D were to have been closed, the results would have been:

		£000
Net profits per Example:		
Department	A	31
	B	32
	C	61
		124
less		
Reallocation of fixed costs		
Department	D	159
Total net profit/(loss)		(35)

On the basis of these figures it is readily apparent that Department D should not be closed.

13.9 Hire or buy

Companies sometimes need an item of plant or equipment, such as a compressor, for a particular job or contract. The question then arises of whether the item should be hired or bought, a decision which can be based on a comparison of the relevant costs of each alternative.

Example

A company needs an item of equipment for a job on which it is engaged, on completion of which it will have no further use for the item. Two possibilities are under consideration:

Possibility 1 To hire the equipment for the estimated 120 days for which it will be required, at a hire rate of £60 per day.

Possibility 2 To buy the equipment for £20 000. Its estimated residual value after the period of use is estimated to be £8 000.

Whether the equipment is hired or bought, the operating cost has been estimated at £12 per day.

Solution

The financial effects of each possibility are tabulated below:

	Possibility 1 (Hire) £	*Possibility 2 (Buy)* £
Sales	*Not applicable*	*Not applicable*
Variable costs		
Hire charge (120 × £60)	7 200	Not applicable
Operating costs (120 × £12)	1 440	1 440
Total	8 640	1 440
Contribution (negative)	(8 640)	(1 440)
Fixed costs		
Net acquisition cost (20 000 – 8 000)	Not applicable	(12 000)
Net cost	(8 640)	(13 440)

Although the buying alternative produces a lower negative contribution of £1 440, indicating a cost avoidance of £7 200 (8 640 – 1 440), when the net cash outflow effect of the acquisition is taken into account, the situation changes. Hiring is seen to be preferable alternative, producing an overall cost saving of £4 800 (13 440 – 8 640) by comparison with buying.

13.10 Make or buy

In a manufacturing company decisions have to be made as to whether to make a particular item or component, or to buy it from another manufacturer. The principles on which the decision is reached are the same as in the hire or buy situation (dealt with in Section 13.9 above): that is, by a comparison of the relevant costs of the alternatives.

Example

A joinery company has received an enquiry for the supply of fitted units in a small housing development. The value of the order would be £30 000. If the order is accepted, it could be fulfilled in one of two ways:

Alternative 1 is to make the units in the company's own workshops, for which the estimated costs would be:

	£
Timber and fittings	10 000
Direct labour cost (cutting, machining, finishing assembling)	5 700
Variable overheads	3 500
Fixed overheads	5 600

Alternative 2 is to buy the units ready-made from another company at an estimated cost of £21 000 and incurring fixed overheads of £2 100.

The choice of modes is to be based on the higher amount of resulting profit.

Solution

The financial effects of each alternative are now tabulated:

	Alternative 1 (Make) £	*Alternative 2 (Buy)* £
Sales	30 000	30 000
less		
Variable costs		
Materials	10 000	21 000
Labour	5 700	Not applicable
Overheads	3 500	Not applicable
Total	19 200	21 000
Contribution	10 800	9 000
less		
Fixed overheads	5 600	2 100
Net profit/(loss)	5 200	6 900

At the contribution stage the 'make' alternative seems the better course of action, but ranking is reversed when the extra fixed overheads are considered. The final position is that the buying option will result in a figure of profit £1 700 greater (6 900 − 5 200) than the other alternative.

13.11 Limiting factor constraints

A further situation for consideration is where a key factor of production is in short supply. The key, or limiting, factor may be machine hours, skilled labour hours, a scarce material, floor area or any other factor which has the effect of restricting the volume of output. Under these circumstances, calculations have to be made to find the production mix needed to achieve the maximum profit.

Example

A joinery company manufacturers tables, chairs and cabinets, data for which are below:

	Tables per unit	*Chairs per set*	*Cabinets per unit*
Machine hours required	4	8	3
	£	£	£
Selling price	80	112	104
Variable costs			
Materials	31	37	64
Labour	20	24	17
Overheads	8	16	6
	No.	*Sets*	*No.*
Orders received	300	500	530

There are 5 145 machine hours available in the period, this being the limiting factor. The problem is to calculate the optimum output mix, that is, the one which will produce the most profit.

Solution

The first step is to tabulate the information:

	Tables per unit £	*Chairs per set* £	*Cabinets per unit* £
Selling price	80	112	104
less			
Variable costs			
Materials	31	37	64
Labour	20	24	17
Overheads	8	16	6
	—	—	—

Table continued over

Table continued	*Tables per unit £*	*Chairs per set £*	*Cabinets per unit £*
Total	59	77	87
Contribution	21	35	17
Ranking	2nd	1st	3rd

On the basis of the above figures, it would seem advisable for the company to concentrate on making sets of chairs, then tables and finally cabinets.

If there are enough machine hours available all the orders can be fulfilled, but if the orders require more than 5 145 machine hours, a priority system will have to be established which, on the above calculations, will coincide with the contribution rankings.

Further calculations are needed to calculate the required machine hours:

	Tables No.	*Chairs sets*	*Cabinets No.*	*Total No.*
Quantity ordered	300	500	530	Not applicable
Machine	Hrs	Hrs	Hrs	Hrs
per unit/set	4	8	3	Not applicable
per order	1 200	4 000	1 590	6 790

The total machine hours needed to manufacture all the quantities ordered are 6 790. This figure is greatly in excess of the maximum available machine hours, therefore some of the orders will have to be rejected.

In this type of situation the technique for maximising profit is to find the contribution ranking in relation to the scarce resource, thus:

	Tables £	*Chairs £*	*Cabinets £*
Contribution per unit/set	21	35	17
	Hrs	Hrs	Hrs
Machine hours required per unit/set	4	8	3
	£	£	£
Contribution per machine hour	5.25	4.375	5.67
Contribution ranking	2nd	3rd	1st

It can be seen that, on this basis, the rankings have altered drastically. Cabinets originally ranked 3rd are now 1st, and Chairs originally ranked 1st are now 3rd. Production quantities can now be determined on the basis of the latest rankings:

	Hours Available hrs	Used hrs	*Ranking*	*Quantity produced units/sets*
	5 145			
		1 590 (530 × 3)	1st cabinets	530
	3 555			
		1 200 (300 × 4)	2nd tables	300
	2 355			
		2 352 (294 × 8)	3rd chairs	294 [2 355/8]
	3			
Total hours used		5 142		

On the basis of these figures, the contribution would amount to:

	Tables No.	*Chairs sets*	*Cabinets No.*	*Total*
Production	300	294	530	
	£	£	£	£
Contribution				
per unit	21	35	17	
amount	6 300	10 290	9 010	25 600

13.12 Relevance of short-term decision-making methods

The methods and techniques illustrated in this chapter enable business decisions of a short-term nature to be reached on the basis of purely financial grounds. The course of action chosen is the one that will produce the most profit or avoid the most cost.

In reality, business decisions are taken after considering many factors, not only financial ones. For example, in a closure situation, explained in Section 13.8, management would take into account the effect on the morale of the workforce; the strength of resistance from the trade unions; possible changed perceptions of the company by its customers and suppliers; adverse publicity in the trade press; and other matters. It must be understood, therefore, that financial considerations are an important, but not necessarily an overriding, factor in short-term decision-making.

14 Long-term Decision-making Techniques

14.1 Introduction

In the introduction to Chapter 13 the need for and circumstances of decision-making were noted and the nature of short-term decisions was explained.

By way of contrast to the subject matter of Chapter 13, this chapter deals with long-term decision-making techniques. They are labelled 'long-term' because the situations with which they are concerned involve very substantial and costly projects expected to operate for at least four years, but probably for even longer. Once the decision has been made it is virtually impossible for the business to alter it in view of the large sums of money invested in it from the outset. Abandonment of the selected project in favour of another could cripple the company financially.

14.2 Common long-term decision situations

There are many situations in this category. A fundamental change is one example. A manufacturing company currently generating its own power by means of a coal-fired system might consider replacing it with a gas-fired or oil-fired system. The company would have to evaluate the financial effects of the alternatives, bearing in mind that there may be several systems suppliers.

Another situation is where a company wants to update or upgrade its production facilities to maintain its competitive advantage by employing the latest technological advances. This might involve replacing the existing plant and machinery with a fully automated, computer-controlled system. After investigation of suitable systems, the directors of the company might have to decide which system to adopt out of three feasible alternatives. Long-term decision-making techniques would give a clear indication as to which alternative should be selected on purely financial grounds. This is known as capital investment appraisal.

14.3 Long-term decision-making methods

Four basic methods are to be explained and illustrated in this chapter. There are other, more sophisticated methods, but these are outside the scope of this book. The methods dealt with in this chapter are:

(a) payback;
(b) accounting rate of return;
(c) net present value (discounted cash flow); and
(d) yield (internal rate of return).

Each of these will be explained and illustrated in the sections which follow.

14.4 Payback

Under this method the decision is based on the *velocity* with which the project pays back. By this is meant the period of time it takes for the project to general a net inflow equivalent to its initial outflow. The project with the shorter, or shortest, payback period is selected, as shown in the following example.

Example

A company is about to replace some plant and is considering the alternatives: Machine A costing £250 000 and Machine B costing £300 000. Cash inflows have been forecast as follows:

Year	*Machine A* *£000*	*Machine B* *£000*
1	*70*	*120*
2	*90*	*180*
3	*90*	*40*
4	*70*	*30*
5	*50*	*30*

The directors have decided to make the choice on the basis of payback.

Solution

The net cash inflows of individual years are accumulated to the point where they equal the initial outflow:

Year	*Machine A*		*Machine B*	
	Current *£000*	*Cumulative* *£000*	*Current* *£000*	*Cumulative* *£000*
1	70	70	120	120
2	90	160	180	300
3	90	250		

Machine B would be chosen because its payback period (two years) is shorter than that of Machine A (three years).

There are various reasons why payback is used. It is easy to calculate and to understand. Where businesses are short of cash funds, as many small and medium-sized businesses are, recouping the initial outlay on an investment as quickly as possible outweighs all other considerations. Bearing in mind that the net cash flow figures, on which the decisions are based, are forecasts, there is a greater likelihood that the figures for the earlier years are more accurate than those of the later years. For this reason there is thought to be less risk attached to the payback method.

Payback does, however, have some weaknesses. It ignores all cash flows arising after payback has been achieved and it gives an equal weighting to the net cash flow of each year, ignoring the timing aspects and the time value of money.

14.5 Accounting rate of return

Under this method, a profitability percentage is calculated for each alternative, for comparison with a predetermined percentage. Any project whose percentage falls below the standard is rejected, and the project with the highest percentage above it is accepted.

Example

The facts are as shown in the payback example in Section 14.4, but further information is available. Both Machine A and Machine B will have a five-year life, with a residual value of £50 000 at the end of it. Depreciation will be charged on a straight-line basis to each of the five years. Profits/(losses) before tax can be arrived at by taking the cash flows in Section 14.4 and adjusting them by the annual depreciation.

The standard against which projects are judged is average annual pre-tax profits reaching a minimum of 10 per cent of initial investment or 20 per cent of average investment.

Solution

	Machine A *£000*	*Machine B* *£000*
Initial investment	250	300
Residual value	(50)	(50)
	——	——

Table continued next page

Table continues	*Machine A* *£000*		*Machine B* *£000*	
Depreciation				
Total	200		250	
Annual (5 years)	40		50	
Net profit/(loss) before tax				
Year				
1	30	[70 – 40]	70	[120 – 50]
2	50	[90 – 40]	130	[180 – 50]
3	50	[90 – 40]	(10)	[40 – 50]
4	30	[70 – 40]	(20)	[30 – 50]
5	10	[50 – 40]	(20)	[30 – 50]
Total	170		150	
Annual average	34	[170/5]	30	[150/5]

The rate of return can be found from one or other of the following formulae:

(1) $$\frac{\text{Annual average profits}}{\text{Initial investment}} \times \frac{100}{1} = \text{return percentage}$$

(2) $$\frac{\text{Annual average profits}}{\text{Average investment}} \times \frac{100}{1} = \text{return percentage}$$

Average investment is taken to mean:

$$\frac{\text{Initial investment plus residual value}}{2}$$

Substituting figures for the items in the above formulae:

	Machine A	*Machine B*
Formula (1) (Standard 10%)	$\frac{34}{250} \times \frac{100}{1} = 13.6\%$	$\frac{30}{300} \times \frac{100}{1} = 10\%$
Formula (2) (Standard 20%)	$\frac{34}{(250+50)/2} \times \frac{100}{1} = 22.7\%$	$\frac{30}{(300+50)/2} \times \frac{100}{1} = 17.1\%$

If the decision is based on Formula (1), both machines are acceptable in that they satisfy the criterion (10 per cent) but Machine A would be selected because of its higher percentage (13.6 per cent).

Using Formula (2), Machine B would be rejected outright as its 17.1 per cent falls short of the minimum 20 per cent, leaving Machine A to be selected.

It is interesting to note that, using the accounting rate of return method, Machine A would be chosen, which is the opposite of the decision reached when payback, which favoured Machine B, was employed.

The percentage used as the yardstick against which project rates of return are compared, is often the actual or the target rate of return on capital employed (ROCE) profitability percentage, calculation of which was described and illustrated in Chapter 8, Section 8.6. If a project were to be accepted with a ROCE below that of the actual or the target, the effect would be to depress these figures. For that reason, the project is rejected as unacceptable. Unlike payback, it takes into account the performance of projects over the whole of their lives. It is, however, easy to calculate and understand and, as stated above, is comparable with ROCE. Disadvantages of the method are that it takes no account of the incidence of the net profits/(losses) or of the time value of money.

14.6 The time value of money

A weakness, already pointed out, of both the payback and accounting rate of return methods of project appraisal is that they both ignore not only the timings of the cash flows and profit before tax but also the time value of money. Methods have been devised to overcome these drawbacks, two of which, net present value and yield, are the subject of Sections 14.7 and 14.8. As a preliminary, however, it is necessary to consider the time value of money, which is the basis for each of these methods.

The expression 'time value of money' recognises the fact that money can be invested to earn interest which, if left without being withdrawn, swells the original sum so that the next calculation is based on the increased figure. It has what might be described as a snowball effect whereby the original sum invested is augmented by ever-increasing amounts as time goes by. This principle is illustrated in Table 14.1, which assumes that £100 is invested at 10 per cent per annum for a period of five years. This rate has been chosen for the illustration because the calculations are easier to follow than would otherwise be the case if a different rate were used.

By the end of Year 5, the original £100 has increased to £161 by the addition of interest to the increasing balance at each year end. This process is termed compounding. In deciding between projects we reverse this process and ask what amount would have to be invested at the start, at a given rate of interest, to produce the net cash flow of any given year of the project. The principles are illustrated in Table 14.2 using the facts of Table 14.1.

Table 14.1 Effect of investment at 10 per cent per annum

				End of year		
	Start	*1*	*2*	*3*	*4*	*5*
	£	£	£	£	£	£
Sum invested	100					
Interest		10				
Balance		110				
Interest			11			
Balance			121			
Interest				12.1		
Balance				133.1		
Interest					13.3	
Balance					146.4	
Interest						14.6
Balance						161.0

Table 14.2 Derivation of initial investment from future cash flows

				End of year		
	Start	*1*	*2*	*3*	*4*	*5*
	£	£	£	£	£	£
Net cash inflow		110				
Interest		10				
Initial investment	100 ←					
Net cash inflow			121			
Interest		10 ←	11			
Initial investment	100 ←					
Net cash inflow				133.1		
Interest		10 ←	11 ←	12.1		
Initial investment	100 ←					
Net cash inflow					146.4	
Interest		10 ←	11 ←	12.1 ←	13.3	
Initial investment	100 ←					
Net cash inflow						161.0
Interest		10 ←	11 ←	12.1 ←	13.3 ←	14.6
Initial investment	100 ←					

Table 14.2 shows how a net cash inflow received five years from the outset requires an initial investment of £100 at 10 per cent per annum. The £100 is described as the present value of £161 receivable in five years' time.

There are short-cut methods for calculating the amounts shown in Table 14.1 and Table 14.2. The formula for finding the balance at the end of a given year from an amount invested at a stated interest percentage is:

$$A = P(1 + r)^t$$

Key:

A = the compounded amount at the end of the given year
P = initial investment
r = interest rate in decimal form
t = number of given years.

In Table 14.1 the balance at the end of Year 5 could have been calculated by substituting figures in the above formula thus:

$$\begin{aligned} A &= £100\,(1 + 1.0)^5 \\ &= £100\,(1.10)^5 \text{ that is, } £100 \times 1.10 \times 1.10 \times 1.10 \times 1.10 \times 1.10 \\ &= £161.05 \end{aligned}$$

Table 14.2 showed the reversal of this process, whereby future cash flows are discounted back to a present value. Here again the figures can be arrived at by applying a formula to find the amount needed to be invested initially at a stated interest percentage to produce a balance at the end of a given year:

$$P = \frac{A}{(1 + r)^t}$$

Key:

P = initial investment
A = the compounded amount at the end of the given year
r = interest rate in decimal form
t = number of given years.

In Table 14.2 the initial investment required to produce the balance at the end of Year 5 could have been calculated by substituting figures in this formula thus:

$$P = \frac{£161.0}{(1 + 0.10)^5}$$

$$= \frac{£161.0}{(1.10)^5}$$

$$= \frac{161.0}{1.10 \times 1.10 \times 1.10 \times 1.10 \times 1.10}$$

$$= £100$$

In other words, £100 is the present value of £161 receivable in five years time, after interest has been compounded at the rate of 10 per cent per annum. The above calculations can be simplified even further by the application of a factor, so that the formula becomes:

$P = A \times$ the factor appropriate to r^t

Using the above figures, the factor would be:

$$\frac{1}{(1.10)^5} = 0.621$$

therefore $P = £161.0 \times 0.621$
$\qquad = £100$

There is no need for the factor to be calculated on each occasion because tables, known as Present Value Tables, are available which give the present value factors from 1 per cent to 20 per cent and thereafter in steps to 50 per cent for each of 20 years. It is a simple matter to select the present value factor, for a particular year and interest rate, and to apply it to a future cash flow to arrive at a present value. Copies of these tables are included in this book in Appendix II on pp. 202 and 203.

Example

Find the present value of:

(a) £1700 receivable in five years' time at 12% p.a.
(b) £1700 receivable in four years' time at 12% p.a.
(c) £1700 receivable in four years' time at 16% p.a.
(d) £ 900 receivable in three years' time at 7% p.a.

Solution

Using the factors obtained from Appendix II, the present values are:

(a) 1700 × 0.567 = £963.90
(b) 1700 × 0.636 = £1081.20
(c) 1700 × 0.552 = £938.40
(d) 900 × 0.816 = £734.40

These principles are applied in the net present value and yield methods of capital investment appraisal which follow in Sections 14.7 and 14.8.

14.7 Net present value (discounted cash flow)

The basis of this method is that, for a project to be acceptable, the sum of the present values of the net cash flows estimated to arise over the life of the project must at least equal the initial outlay. If it fails to do so, the project would be rejected automatically, on the grounds that, after eliminating interest receivable, it was unable to generate a net cash inflow sufficient to cover its original outlay. Where some or all of the projects produce present values in excess of the initial outlays, the project with the higher or highest excess, termed a *positive net present value*, is chosen. If the projects being compared require different amounts of initial investment outlay, selections based on absolute figures of net present values would be invalid; present value indices are used instead. Choice of interest rates used in the calculations is dealt with in Section 14.9. These points are illustrated in the example which follows.

Example

A company is in process of deciding which of two items of manufacturing plant to install. For this purpose, an interest rate of 12 per cent per annum is the criterion. This further information is available:

		Plant X *£000*	*Plant Y* *£000*
Initial outlay		500	700
Net cash inflows			
	Year		
	1	80	250
	2	110	270
	3	190	200
	4	210	160
	5	130	110
Sale proceeds of plant	5	40	70

The selection is to be based on net present value.

Solution

The first step is to produce a table in which the present value factors for 12 per cent for each of the years 1 to 5 are applied to the net cash inflows to produce present values (discounted cash flows), which are then aggregated and compared with the initial outlays to disclose a net present value:

		Machine X		*Machine Y*	
Year	*Present value factor (12%)*	*Net cash inflow £000*	*Present value £000*	*Net cash inflow £000*	*Present value £000*
1	0.893	80	71.44	250	223.25
2	0.797	110	87.67	270	215.19
3	0.712	190	135.28	200	142.40
4	0.636	210	133.56	160	101.76
5	0.567	130	73.71	110	62.37
5	0.567	40	22.68	70	39.69
Present value			524.34		784.66
less					
Initial outlay			500.00		700.00
Net present value			24.34		84.66

Both machines meet the 12 per cent criterion, as is shown by the fact that their net present values are positive. At first sight it would seem that Machine Y should be selected, on the basis of its higher net present value. However, in instances where the initial outlays are different, the absolute figures do not necessarily provide the correct solution. This is decided by an index number calculated as shown below:

$$\frac{\text{Present value}}{\text{Initial outlay}} = \text{Net present value (NPV) index}$$

The index numbers in this example can be calculated as:

$$\text{Machine X} \quad \frac{524.34}{500.00} = 1.05$$

$$\text{Machine Y} \quad \frac{784.66}{700.00} = 1.12$$

Machine Y would be selected on the basis of its higher NPV index.

14.8 Yield (internal rate of return)

The net present value method of capital investment appraisal indicates clearly whether the projects under consideration generate net cash flows in excess of a specified interest rate criterion. A positive NPV indicates that they do, a negative NPV that they do not. In neither case do the calculations show what the actual rate of return of the net cash flows is, but merely whether or not the specified rate has been achieved. The yield method overcomes this problem by calculating the internal rate of return of each project under consideration.

It achieves this objective by one of two routes: either by calculations which are then transferred to a graph and the yield found by interpolation; or by calculation. In either case, the initial calculations are the same.

Two interest rates are selected, one low and the other high, with a wide interval between them. If they have been selected properly, when the NPV calculations have been carried out it will be found that the low interest rate has produced a positive NPV and the high rate a negative NPV. The yield must lie somewhere between these two extremes and is found by plotting the two NPVs on a graph, joining them with a straight line and arriving at the yield. By interpolation, the point of intersection on the interest axis is the yield percentage. It is usual for the vertical axis to show the net present values and the horizontal axis the interest rates. This should be clear from Figure 14.1.

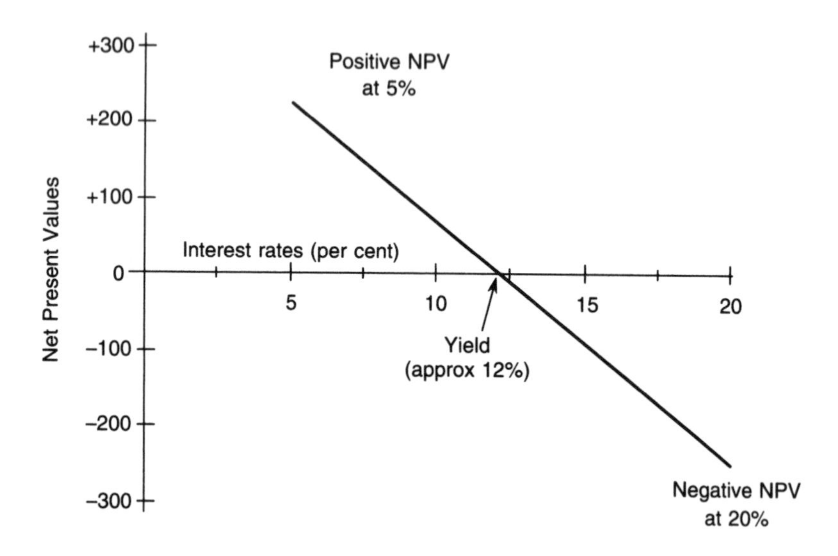

Figure 14.1 Example of a yield graph

A full example now follows:

Example

The facts are the same as in Section 14.7. Management needs to know the yield percentage of Machine Y.

Solution

The calculations in 14.7 were based on an interest rate of 12 per cent and in the case of Machine Y produced a positive NPV of £84 660, thus indicating that the actual yield must be at a rate higher than 12 per cent. A low rate and a high rate in relation to 12 per cent would be 5 per cent and 30 per cent and there is a wide gap between these figures. The start point, therefore, is to perform present value calculations on the net cash inflows of Machine Y separately for the 5 per cent and 30 per cent rates:

			Machine Y		
Year	*Net cash inflow £000*	*Present value factor (5%)*	*Present value £000*	*Present value factor (30%)*	*Present value £000*
1	250	0.952	238.00	0.769	192.25
2	270	0.907	244.89	0.592	159.84
3	200	0.864	172.80	0.455	91.00
4	160	0.823	131.68	0.350	56.00
5	110	0.784	86.24	0.269	29.59
5	70	0.784	54.88	0.269	18.83
Present value			928.49		547.51
less					
Initial outlay			700.00		700.00
Net present value (NPV)			228.49		(152.49)

These figures are now plotted on a graph, shown in Figure 14.2. Against the vertical axis the positive net present values appear above the zero, and negative net present values below it. The horizontal axis starts from the zero and records the interest rates.

After the positive NPV 228 has been plotted at the 5 per cent rate and the negative NPV 152 at the 30 per cent rate, the two plots are joined by a straight line. As can be seen, this line intersects the interest line at 20 per cent precisely, indicating that this is the yield percentage. This gives another insight into the nature of yield which can now be seen to be that percentage which results in a NPV of zero.

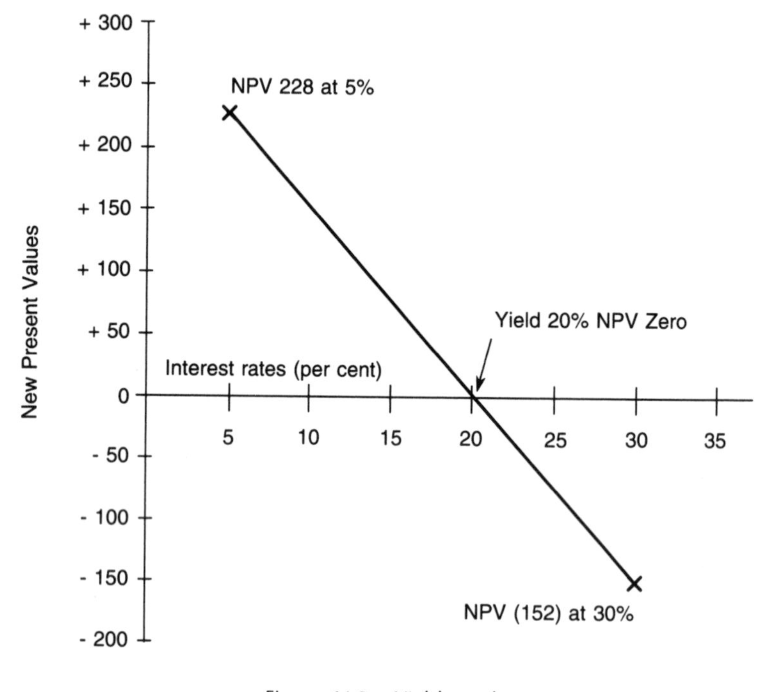

Figure 14.2 Yield graph

It was stated earlier that yield could be found by calculation only. The initial calculations are identical with those in the table of figures preceding the graph, whereby NPVs of 228 and 152 (negative) resulted from 5 per cent and 30 per cent respectively. The yield percentage can be found by applying the formula:

$$A + \left(\frac{(B - A)}{1} \times \frac{a}{(a + b)}\right)$$

Key:

A = interest rate of the positive NPV
B = interest rate of the negative NPV
a = the positive NPV
b = the negative NPV
$a + b$ = the interval between the positive and negative NPVs

Substituting figures in the above formula:

$$5 + \left(\frac{(30 - 5)}{1} \times \frac{228}{(228 + 152)}\right)$$

$$= 5 + \left(\frac{25}{1} \times \frac{228}{380}\right)$$
$$= 5 + (25 \times 0.6)$$
$$= 5 + 15$$
$$= 20\%$$

This produces the same answer as the graph.

14.9 Choice of interest rates

In the NPV method of capital investment appraisal discussed in Section 14.7, calculations were performed on the basis of given interest rates. Under the yield method described in Section 14.8, the percentage was arrived at by one of two means. In neither case was mention made of what the NPV rate represented, or the yield rate measured against for acceptability. These matters are now considered.

Choice of interest rate will depend on the particular circumstances of a company and may be geared to external interest rates or an internal interest rate. If a company had to borrow money to finance the project to be undertaken, it might use as its criterion, for NPV and yield purposes, the rate of interest at which it will have to borrow. On the other hand, a company financing the project from internally generated retained profits might select a rate equal to the rate it could obtain if it were to invest the money outside the business. Alternatively the company could use an internal rate, of which there are two in common use from which a choice could be made. The first is the company's profitability percentage, the return on capital employed (ROCE). The use of this as a yardstick has already been mentioned in Section 14.5 in connection with the accounting rate of return method. Comments made at that point are equally valid here.

In place of the ROCE percentage rate a rate can be used which reflects what is known as the company's cost of capital or, more usually, weighted average cost of capital (WACC), which will now be explained.

Capital employed by a company consists of both share capital and loan capital, as described in Chapter 2, Sections 2.4 and 2.5. Shareholders receive dividends calculated on their shareholdings, and suppliers of loan capital (debenture holders, for example) are remunerated by fixed interest rate payments. The dividends and the interest are regarded as the cost of the continuing availability of the capital. A percentage rate, known as the weighted average cost of capital (WACC) can be calculated on the basis of the yields of the dividends and interest paid. This is illustrated in Table 14.3.

Table 14.3 Calculation of weighted average cost of capital

Capital employed	*Amount* (£)	*Dividend/interest yield rate* (%)	*Dividend/interest per annum* (£)
Share capital			
Ordinary	3 000 000	6	180 000
7% preference	500 000	7	35 000
Loan capital			
9% debentures	700 000	9	63 000
	4 200 000		278 000

Note: Weighted average cost of capital is 6.6 per cent [(278 000 × 100)/4 200 000].

15 Budgetary Control

15.1 Introduction

Efficient businesses, irrespective of size, plan their operations period by period, in advance, and express then in financial terms. Such plans are termed *budgets*; they are the financial targets which the business intends to achieve. Actual results, as they become available, are compared with the budgeted plans and divergences are identified, their causes are investigated and executive action is taken to bring results back into line with the budget. This system is known as *budgetary control* and is analogous to the navigational operations in keeping a ship on course for its destination.

Budgets are usually prepared to span a complete financial year. At any one time a business may have annual budgets for each of the following fifteen years, maintained on a rolling basis; that is, at the expiry of one year the fourteen remaining budgets are updated and revised in the light of the most recently available information, and a new fifteenth budget added.

15.2 Types of budget

It is essential that budgets are prepared in such a way that they are financial expressions of a coherent overall business plan. This objective is usually achieved by preparing *functional budgets*. In a manufacturing concern this means the functions identified in Chapter 9, Section 9.4:

- production;
- selling;
- distribution; and
- administration.

These functional budgets have to interlock and be compatible, one with another. The sales budget, for example, has to be such that the sales volumes are within the capacity of the production budget, after allowing for any surplus sales to be met from stocks in hand or any surplus production to be held in stock.

It is a feature of budgetary control that the budgets should be achievable under realistic conditions rather than under unattainable ideal conditions. For this reason they are compiled on the basis of a joint effort by the functional managers, whose actual performance is to be measured against their budget. Safeguards are in place so that the functional managers do not set levels too low, which they can surpass easily and so enhance their performance-related pay. Budget levels have to be compatible with the overall objectives laid down by the board of directors, or an acceptable explanation given as to why they are not.

15.3 The budget procedure

It is obvious that the preparation of the budget is a lengthy, time-consuming operation that has to be carried out well in advance of the period to which it relates. The start point is for the directors to specify their objectives and the means of achieving them. For example, the directors may decide to expand the business from the start of the next budget period, but to finance the extra investment from existing and future retained profits instead of securing external long term finance.

They may therefore specify the amount of profit which the business needs to generate in the next budget period to meet this objective and, at the same time, be sufficiently large to counteract the eroding effects of inflation and also to maintain or increase dividend distributions to shareholders. Having determined the amount of budgeted profit, the next step is to plan the volume of sales needed to generate it. Production quantities and values can then be budgeted. In conjunction with finished goods stocks, this ensures that the sales budget can be met.

Further budgets are then prepared which stem from the main sales and production budgets. Assuming that all or most of the sales are on credit, a debtors' budget would be needed. The sales budget itself could be analysed by product and by geographical destination. Marketing expenses, including advertising and publicity expenses, would also be budgeted.

For the production function, several budgets would be prepared concerning raw materials – purchases, stocks and usage. The raw materials purchases budget would itself give rise, in conjunction with other items, to a trade creditors' budget. The various quantities and prices of raw materials needed for the budgeted level of production would be encompassed within these budgets.

Conversion of the raw materials into the finished product is effected by direct labour in conjunction with machinery. As a consequence, the direct labour cost budget is prepared by taking account of the direct labour hours of the different grades of labour – unskilled, semi-skilled and skilled, each of which grade is remunerated at a different rate.

Other budgets include individual expense budgets for those items which form part of the variable and fixed overhead expenses and, very importantly, a cash budget.

Using the many individual budgets referred to above, a master budget is compiled to give an overall picture. The master budget consists of a budgeted profit and loss account, a budgeted balance sheet and a budgeted cash flow statement.

15.4 Budget preparation

As has already been stated, the preparation of a master budget from the subsidiary budgets described in Section 15.3 is a lengthy and complex operation, and within the confines of a textbook it is difficult to illustrate the procedure in a manner which adequately reflects these complexities.

The example which now follows attempts to do this, while at the same time keeping the various schedules and workings within manageably comprehensible limits.

Example

The directors of a manufacturing company have decided that the main objective, to be reflected in the budget for Year 7, is to reduce the very sizeable bank overdraft to a maximum level of £350 000. This is to be achieved by a combination of reducing stocks of finished goods, restricting output and increasing long-term borrowing.

On this basis, the following budgeted figures were set:

Production: 20 000 units

Direct materials

Each unit of production requires:
4 units of Material X at £5.00 per unit
7 units of Material Y at £3.00 per unit

Opening stock

Material X – 10 000 units at £5.00 per unit
Material Y – 12 000 units at £3.00 per unit

Closing stock

Material X – 15 000 units at £5.00 per unit
Material Y – 8 000 units at £3.00 per unit

Raw materials creditors:	£
Opening creditors	
Material X	30 000
Material Y	39 000
	69 000
Closing creditors	
Material X	40 000
Material Y	60 000
	100 000

Purchases of raw materials and payments to the creditors are to be derived.

Direct labour

Each unit of production requires:
2 hours of Grade A labour at £5.50 per hour
3 hours of Grade B labour at £6.00 per hour

Machine hours

3.5 machine hours per unit of production

Production overheads

	£	
Variable – per machine hour	4.00	per hour
Fixed (excluding depreciation)	6 000	

Other overheads (fixed, excluding depreciation):

	£
Distribution	£60 000
Administration	£94 000

Fixed assets:

	Production *£000*	*Distribution* *£000*	*Administration* *£000*
End of Year 6:			
Cost	2 877	370	223
Accumulated depreciation	716	63	41
Year 7 additions:			
Cost	443	40	17

Depreciation

Calculated at 10% of cost at the end of the year

Finished goods

Opening stock – 15 000 units at £84.00 per unit
Closing stock – 7 000 units at £84.00 per unit

Trade debtors (for sales):

	£
Opening	470 000
Closing	561 000

Sales

	£	
Selling price	117.00	per unit

Sales and receipts from trade debtors are to be derived

Bank and cash

	£
Closing balance – Year 6 (overdraft)	1 525 000
Receipts in Year 7:	

Trade debtors	To be derived
Issue of further 5% debentures	500 000
Payments in Year 7	
Trade creditors (raw materials)	To be derived
(other)	4 000
Interest (re: Year 6)	50 000
Taxation (re: Year 6	85 000
Dividends (re: Year 6)	60 000
Wages	554 000
Overheads Variable	270 000
Fixed	5 000
Distribution costs	54 000
Administrative expenses	96 000
Fixed asset acquisitions	480 000
Other profit and loss account items in Year 7:	
Interest payable	175 000
Taxation payable on profit	79 000
Proposed dividend	80 000

Other balance sheet items not previously given:

	At 31 December	
	Year 6	*Year 7*
	£000	*£000*
Other debtors	25	25
Other creditors	30	67
Interest payable	50	
Taxation payable	85	20
Proposed dividends	60	
5% debentures	1 000	
Called-up share capital	1 500	
Profit and loss account	172	

Example:

From the foregoing information prepare a master budget for Year 7 comprising:

- profit and loss account;
- balance sheet; and
- cash flow statement.

together with subsidiary budgets, as appropriate.

Solution:

Budgeted profit and loss account for the year ended 31 December Year 7

Budget No.		*£000*	*£000*
1	*Sales*		3 276
	Production cost		
2	Direct materials	820	
3	Direct labour	580	
	Prime cost	1 400	
	Production overheads		
4	Variable	280	
		1 680	
5	Fixed (depreciation)	332	
given	(other)	6	
	Cost of goods manufactured	2 018	
	Finished goods		
1	Opening stock	1 260	
		3 278	
1	Closing stock	588	
	Cost of sales		2 690
	Gross profit		586
6	Distribution costs	101	
6	Administrative expenses	118	
			219
	Trading profit		367
given	Interest payable		175
	Net profit before tax		192
given	Taxation		79
	Net profit after tax		113
given	Dividends proposed		80
	Retained profit		
	For the year		33
given	Brought forward		172
	Carried forward		205

Budgeted balance sheet as at 31 December

Budget No.	*Year 6* £000	£000		*Year 7* £000	£000
			Fixed assets		
5		2650	Tangible		2753
			Current assets		
			Stocks		
7	86		Raw materials	99	
1	1260		Finished goods	588	
			Debtors		
given	470		Trade	561	
given	25		Other	25	
	1841			1273	
			Creditors due in less than one year		
10	1525		Bank overdraft	300	
			Creditors		
given	69		Trade	100	
given	30		Other	67	
given	50		Interest	175	
given	85		Taxation [79 + 20]	99	
given	60		Proposed dividends	80	
	1819			821	
		22	*Net current assets*		452
		2672	*Total assets less current liabilities*		3205
			Creditors due in more than one year		
given		1000	5% debentures [1000 + 500]		1500
		1672			1705
			Share capital and reserves		
given		1500	*Called-up share capital*		1500
			Reserves		
given; P & L		172	Profit and loss account		205
		1672			1705

NB The Year 6 columns would not normally be shown, but have been included here to facilitate the preparation of the budgeted cash flow statement which now follows.

Budgeted cash flow statement for the year ended 31 December Year 7

Budget No.		*£000*	*£000*
P & L	Trading profit	367	
5	Depreciation	397	
	Decrease in stocks [1346 – 687]	659	
	Increase in debtors [561 – 470]	(91)	
	Increase in creditors [167 – 99]	68	
	Net cash inflow from operating activities		1 400
	Returns on investments and servicing of finance		
given	Dividends paid	(60)	
given	Interest paid	(50)	
	Net cash outflow from returns on investments and servicing of finance		(110)
	Taxation		
given	Tax paid		(85)
	Investing activities		
given	Purchase of tangible fixed assets	(480)	
	Net cash outflow from investing activities		(480)
	Net cash inflow before financing		725
	Financing		
given	Issue of 5% debentures	500	
	Net cash inflow from financing		500
	Net increase in cash and cash equivalents		1 225

It can be seen that this reflects the reduction in the bank overdraft:

	£000
Bank overdraft at 31 December	
Year 6	1 525
Year 7	300
Reduction during Year 7	1 225

Thus all the objectives for the Year 7 budget have been achieved:

1. The overdraft has been reduced to below £350 000.
2. Stocks of finished goods have been reduced from £1 260 000 to £588 000.
3. Output of 20 000 units is less than the sales volume, as evidenced by the de-stocking.
4. Long-term borrowing has been increased by £500 000 to £1 500 000.

Individual subsidiary budgets from which the master budget is compiled, and to which they are referenced by number, now follow.

Subsidiary budgets

1. *Sales*

	Units		£
Production	20 000		
Finished goods			
Opening stocks	15 000	at £84 per unit =	1 260 000
Closing stocks	(7 000)	at £84 per unit =	588 000
Sales	28 000	at £117 per unit =	3 276 000

2. *Direct material usage*

	£
Material X	
4 × 20 000 × £5.00	400 000
Material Y	
7 × 20 000 × £3.00	420 000
Total	820 000

3. *Direct labour cost*

	£
Grade A	
2 × 20 000 × £5.50	220 000
Grade B	
3 × 20 000 × £6.00	360 000
Total	580 000

4. *Production overheads – variable*

3.5 × 20 000 × £4.00	280 000

5. *Fixed assets and depreciation*

	Production £000	Distribution £000	Administration £000	Total £000
Cost				
End of Year 6	2 877	370	223	3 470
Year 7 additions	443	40	17	500
End of Year 7	3 320	410	240	3 970
Accumulated depreciation				
End of Year 6	716	63	41	820
Year 7 provision (10%)	332	41	24	397
End of Year 7	1 048	104	65	1 217
Net book value				
End of Year 6	2 161	307	182	2 650
End of Year 7	2 272	306	175	2 753

6. *Distribution costs and administrative expenses*

	Distribution £000	Administration £000
Given	60	94
Depreciation (Budget 5)	41	24
Total	101	118

7. *Raw materials stocks*

		Opening stock £000		Closing stock £000
Material X				
	10 000 units £5.00	50	15 000 × £5.00	75
Material Y				
	12 000 units £3.00	36	8 000 × £3.00	24
Total		86		99

8. *Raw materials purchases and payments to creditors*

	Units No.	*Price per unit* £	*Amount* £000
Material X			
Closing stock	15 000	5.00	75
Usage (budget 2)	80 000	5.00	400
Opening stock	(10 000)	5.00	(50)
Purchases	85 000	5.00	425
Opening creditors			30
Closing creditors			(40)
Payments to creditors			415
Material Y			
Closing stock	8 000	3.00	24
Usage (budget 2)	140 000	3.00	420
Opening stock	(12 000)	3.00	(36)
Purchases	136 000	3.00	408
Opening creditors			39
Closing creditors			(60)
Payments to creditors			387

	Raw material purchases £000	*Payments to creditors* £000
Material X	425	415
Material Y	408	387
Total	833	802

9. *Receipts from trade debtors*

	£000
Opening debtors	470
Sales (budget)	3 276
Closing debtors	(561)
Receipts from trade debtors	3 185

10. *Bank and cash*	*£000*
Receipts from:	
Trade debtors (Budget 9)	3 185
Issue of 5% debentures (given)	500
Total receipts	3 685
Payments to:	
Trade creditors (Budget 8)	802
Other creditors (given)	4
Interest (given)	50
Taxation (given)	85
Dividends (given)	60
Wages (given)	554
Production overheads variable (given)	270
fixed (given)	5
Distribution costs (given)	54
Administrative expenses (given)	96
Fixed assets (given)	480
Total payments	2 460
Excess of receipts over payments	1 225
Opening overdraft	(1 525)
Closing overdraft	(300)

15.5 Budget comparison

When at the conclusion of the accounting period for which the budget has been prepared the actual results are available, they are compared with the budget. This comparison is usually in tabular format. In the following example, the 'cost of goods manufactured' figures from the budgeted profit and loss account in Section 15.4 are the basis for comparison with the actual figures.

Example

For Year 7 the budgeted production was 20 000 units, the production cost of which was as given in Section 15.4. Actual production for the same period was 25 000 units, at a cost detailed in the comparison statement. Variances of actual figures from the budgeted are favourable (F) if they represent a cost saving or adverse (A) if they represent an overspend.

Budget/actual comparisons – Year 7

	Budget	*Actual*	*Variance*
Production units	20 000	25 000	+5 000
	£000	£000	£000
Production costs			
Direct materials	820	950	130 (A)
Direct labour	580	730	150 (A)
Prime cost	1 400	1 680	280 (A)
Production overheads			
Variable	280	330	50 (A)
	1 680	2 010	330 (A)
Fixed (depreciation)	332	347	15 (A)
(other)	6	15	9 (A)
Cost of goods manufactured	2 018	2 372	354 (A)

On a purely arithmetical basis the company's actual overall cost of production has exceeded budget by £354 000, with adverse variances for each cost item. This must be viewed with caution. Actual production was 5 000 units above budget; this fact alone would cause an increase in actual costs. As the statement stands at the moment, it is not apparent how much of the increases is due to the increased volume of production and how much to other causes. This lack of clarity is regarded as a defect of such importance that special measures need to be taken to remedy it. The solution is for the actual figures to be compared, not to the original budget, but to a flexed version of it, the effect of which is to remove those parts of the variances attributable to the volume variation. Variances then remaining are attributable to other causes.

15.6 Flexible budget preparation

The need for flexible budgets has been explained. Preparation of flexible budgets is a comparatively straightforward operation and does not require full calculations of the type illustrated in Section 15.4. Instead, the original budget is the start point for adjusting the figures. The individual items in the budget have first to be identified as either fixed or variable. Analysis into these categories was dealt with in Chapter 9. By definition, the direct materials and direct labour costs are variable items. Production overheads are a mixture, from which the fixed element has to be siphoned off from the variable. In some cases, this will have been done during the budget preparation; in other cases it will not. Where it is not apparent whether an item is fixed or variable, or a combination of the two (a semi-variable), it is relatively easy to establish its behaviour.

Example

The production and other costs at three different levels of output have been ascertained, as shown below.

Production (000 units)	*60*	*70*	*80*
	£000	*£000*	*£000*
Costs			
Materials	600	700	800
Labour	900	1 050	1 200
Production overheads	780	860	940
Other overheads	500	500	500
	2 780	3 110	3 440

A cursory look reveals that other overheads are a fixed cost because they remain at a constant £500 000 over each output level. The task of identifying the variable elements requires the figures to be analysed on a 'per unit' basis.

Production (000 units)	*60*		*70*		*80*	
	Per unit		*Per unit*		*Per unit*	
	£	*£000*	*£*	*£000*	*£*	*£000*
Costs						
Materials	10	600	10	700	10	800
Labour	15	900	15	1 050	15	1 200
Production overheads	13	780	12.3	860	11.8	940
Other overheads	8.3	500	7.1	500	6.2	500
	46.3	2 780	44.4	3 110	43.0	3 440

It can now be seen that, irrespective of the output level, the cost per unit of materials and of labour has remained constant at £10 per unit and £15 per unit, respectively. This fact clearly indicates that these are both variable costs.

Production overheads pose a problem. They are not fixed costs because the amount alters at each output level, rising from £780 000 to £940 000. Nor can they be regarded as variable costs because the cost per unit alters at each output level, reducing from £13 per unit down to £11.8 per unit. A variable cost would have displayed a constant unit cost. Production overheads must therefore be a semi-variable cost: that is, they are composed of a mixture of fixed and variable elements.

The problem then is to devise some means of segregating the two elements. This can be done by further analysis, involving the measurement of the movement in the amount of production overheads between levels.

		Difference			*Difference*		
Production (000 units)	60		+10	70		+10	80
		Per unit			*Per unit*		
	£000	*£*	*£000*	*£000*	*£*	*£000*	*£000*
Production overheads	780	8	+80	860	8	+80	940

Between the output levels there is a difference of 10 000 units and a cost of £80 000, giving a constant cost per unit of £8. This represents the variable element, which can now be isolated.

Production (000 units)	*60*	*70*	*80*
	£000	£000	£000
Production overheads	780	860	940
less			
Variable element	480	560	640
at (£8 per unit)			
Fixed element	300	300	300

It is self-evident that the fixed element of the semi-variable production overheads has been identified correctly, because the amount is a constant £300 000 at each level. Using the information obtained so far, it is now possible to construct a flexible budget for any other output level.

Example

The facts are as in the previous example. It is now necessary to prepare a budget for an output of 74 000 units.

Solution

The budget is prepared using the cost per unit for all the variable items and the static amounts for the fixed overheads.

Flexed Budget

Production (000 units)	74	
	Per unit £	£000
Costs		
Materials	10	740
Labour	15	1 110
Production overheads		
Variable	8	592
Fixed		300
Other overheads		500
		3 242

Returning now to the budget/actual comparison given in Section 15.5, the original budget, based on an output of 20 000 units, can be flexed to the actual output of 25 000 units. On this occasion there is no need to identify the fixed and variable elements, because this information has been given – the direct labour, direct materials and part of the production overheads are all variable, and the remaining items are fixed, thus:

Flexed Budget – Year 7

Production (000) units	20		25	
	Per unit £	£000	*Per unit* £	£000
Production cost				
Direct materials	41	820	41	1 025
Direct labour	29	580	29	725
Prime cost	70	1 400	70	1 750
Production overheads				
Variable	14	280	14	350
		1 680		2 100

Fixed (Depreciation)	332	332
(Other)	6	6
Cost of goods manufactured	2018	2438

The original budget, flexed to an output level of 25 000 units, can now be used as the standard against which actual results are compared as is now shown.

Flexed budget/actual comparison – Year 7

	Flexed budget	*Actual*	*Variance*
Production (000) units	25 000	25 000	–
	£000	£000	£000
Production cost			
Direct materials	1025	950	75 (F)
Direct labour	725	730	5 (A)
Prime cost	1750	1680	70 (F)
Production overheads			
Variable	350	330	20 (F)
	2 100	2 010	90 (F)
Fixed (Depreciation)	332	347	15 (A)
(Other)	6	15	9 (A)
Cost of goods manufactured	2 438	2 372	66 (F)

A totally different picture has now emerged from that disclosed in Section 15.5. Then there was an overall adverse variance of £354 000, now there is a favourable variance of £66 000, with consequent changes on individual items. All the variable costs, with one minor exception, disclose favourable variances. Now that the flexible budget has eliminated differences attributable to the increased volume of production, those remaining have other causes. Further investigations would be instigated into the reasons behind these variances. It is important that all significant variances be investigated whether adverse or not. Reasons behind an adverse variance are important because they indicate measures which can be taken to overcome the problem. Of no lesser importance is the examination into the causes of the favourable variances – they give an insight into economies and efficiencies which can be harnessed for the benefit of the business in the future.

15.7 Frequency of comparisons

Up to this point of the chapter the actual/budget comparisons have been explained and illustrated on an actual basis. If this were the only time the comparisons were carried out, the system would be seriously flawed. The reason for stating this is that, if comparisons were only made after a year had elapsed, a whole year's variances would have accumulated before these were ascertained. As a consequence, in the meantime, no corrective action could be taken to eradicate the causes of adverse variances or to harness the benefits indicated by the favourable variances.

For control purposes, therefore, the budget for the current year is phased to quarters and months of the year. Some significant items may also be phased to weeks of the month. It is this phasing which enables budgetary control to be so effective. The sooner any variances can be identified, the sooner can corrective action be initiated, therefore actuals are compared, as soon as they are available, with their budgetary counterpart. Depending on the phasing, this generally means a few days after the end of the month or week. In this way, the time lag between an actual occurrence and its comparison with budget is kept to an absolute minimum.

16 Cash Budgeting

16.1 Introduction

Effective management of the cash resources of a business is critical to its survival. Various aspects of control of cash and cash equivalents were covered in Chapter 6, which dealt with cash flow statements, and Chapter 8, where the monitoring of liquidity by ratio analysis was covered. In Chapter 15, Section 15.4 a budget (Number 10) was prepared for cash and bank transactions.

The fundamental shortcoming of figures prepared on an annual basis, as was the case with all the examples quoted in the previous paragraph, is that they disclose the position at the beginning and end of the year but not the position in intervening periods. Thus a business may have adequate bank and cash balances to start and finish the period but in the gap between the two it may run out of cash completely and be in overdraft. An annual budget would not detect this situation. It is important, therefore, that the annual cash budget be phased down to months (or weeks). If it then becomes apparent that the business is going to be short of cash, steps can be taken well in advance to raise cash or to obtain an authorised overdraft.

16.2 Phased cash budget preparation

This involves determining the incidence of each item of receipts and payments. The resulting figures are recorded on a grid schedule on which the closing balances, positive or negative (overdraft) are then derived. In Chapter 15, Section 15.4, a master budget was prepared for Year 7, supported by subsidiary budgets. One of these, Number 10, was a bank and cash budget. This annual budget will now be phased to months in the example below.

Example

The facts are as in supplementary budget Number 10 in Chapter 15, Section 15.4. This additional information is available:

1. Receipts from trade debtors – £220 000 in January, £225 000 per month from February to May, and £295 000 per month from June onwards.
2. The issue of additional 5 per cent debentures (£500 000) would be made in November.

3. Payments to trade creditors – £69 000 in January, £73 000 in February, £70 000 in both March and April, and £65 000 per month from May onwards.
4. Other creditors (£4 000) to be paid in May.
5. Interest to be paid in January (£50 000)
6. Payments of taxation liabilities – £25 000 in July and £60 000 in September.
7. Dividends (£60 000) to be paid in April.
8. Payments of wages – £44 000 per month from January to August, £50 000 in both September and October, and £51 000 in both November and December.
9. Payments of overhead expenses – variable £22 000 per month from January to June, and £23 000 per month from July to December; fixed (£5 000) payable in February.
10. Payments of distribution costs – £4 000 per month from January to June; £5 000 per month from July to December.
11. Payments of administrative expenses £8 000 per month from January to December.
12. Payments for fixed assets (£480 000) to be in December.

Required

Prepare the phased cash budget for Year 7. The solution is shown in Table 16.1, on the following page.

In fact, the phasing of a cash budget can be a more complicated operation than in the example just given, particularly for a manufacturing company. This is illustrated in a second example.

Example

The following information is available for the Year 4 budget:

1.

	May	*June*	*July*	*Aug*	*Sept*	*Oct*	*Nov*	*Dec*
Quantities								
Sales (units)								
cash	20	20	30	30	40	40	40	50
on credit	240	280	270	310	320	330	360	370
Production (units)	260	260	260	340	340	340	340	380

2. Sales prices will be £50 per unit from May to July, £60 from August to November, and £70 in December. Buyers on credit will be granted two months' credit: that is, goods sold to them in May will have to be paid for in July.
3. Production costs throughout the period will be labour £10, overheads £7 and materials £18, in each case per unit produced. Labour and overhead costs will be paid in the month in which they are incurred.

   Material purchases will be paid one month after the materials have been bought (see item 4 below).

Table 16.1 Phased cash budget – Year 7

	Jan	*Feb*	*Mar*	*Apr*	*May*	*June*	*July*	*Aug*	*Sept*	*Oct*	*Nov*	*Dec*	*Total*
Receipts													
Trade debtors	220	225	225	225	225	295	295	295	295	295	295	295	3 185
5% debentures											500		500
Total (R)	220	225	225	225	225	295	295	295	295	295	795	295	3 685
Payments													
Creditors													
Trade	69	73	70	70	65	65	65	65	65	65	65	65	802
Other				4									4
Interest	50												50
Taxation							25		60				85
Dividends				60									60
Wages	44	44	44	44	44	44	44	44	50	50	51	51	554
Overheads													
Variable	22	22	22	22	22	22	23	23	23	23	23	23	270
Fixed		5											5
Distribution costs	4	4	4	4	4	4	5	5	5	5	5	5	54
Administration expenses	8	8	8	8	8	8	8	8	8	8	8	8	96
Fixed assets												480	480
Total (P)	197	156	148	208	147	143	170	145	211	151	152	632	2 460
(R) – (P)	23	69	77	17	78	152	125	150	84	144	643	(337)	1 225
Opening balance	(1525)	(1 502)	(1 433)	(1 356)	(1 339)	(1 261)	(1 109)	(984)	(834)	(750)	(606)	37	(1 525)
Closing balance	(1 502)	(1 433)	(1 356)	(1 339)	(1 261)	(1 109)	(984)	(834)	(750)	(606)	37	(300)	(300)

Note: Figures in brackets indicate overdraft.

4.

	May	June	July	Aug	Sept	Oct	Nov	Dec
	£000	£000	£000	£000	£000	£000	£000	£000
Materials purchases	5.5	5.5	6	6	6	6	7	7

Creditors for May supplies will be paid in June (see item 3 above).

5. Other payments include:

	£	*Month*
Dividends paid	20 000	July
Fixed assets – vehicles	15 000	August
Taxation	21 000	September

6. Dividends will be received in October (£6 000)

7. The opening bank and cash balance for July will be £14 680.

Required

Prepare the phased cash budget for the six months July to December, Year 4.

Solution

Various workings have to be prepared before the phased cash budget can be compiled.

Workings

	May	*June*	*July*	*Aug.*	*Sept*	*Oct*	*Nov*	*Dec*
Sales and sales receipts								
Cash sales (units)	Not relevant		30	30	40	40	40	50
Price per unit (£)			50	60	60	60	60	70
Cash sales receipts (£000)			1.50	1.80	2.40	2.40	2.40	3.50
Credit sales (units)	240	280	270	310	320	330	360	370
Price per unit (£)	50	50	50	60	60	60	60	70
Credit sales value (£000)	12.00	14.00	13.50	18.60	19.20	19.80	21.60	25.90
Receipts from debtors (£000) (2 months' credit)	→	→	12.00	14.00	13.50	18.60	19.20	19.80

(2) Production cost payments	*May*	*June*	*July*	*Aug*	*Sept*	*Oct*	*Nov*	*Dec*
Production (units)	Not relevant		260	340	340	340	340	380
Labour cost (£000) (£10 per unit)			2.60	3.40	3.40	3.40	3.40	3.80
Overhead cost (£000) (£7 per unit)			1.82	2.38	2.38	2.38	2.38	2.66
Materials (£000) *purchases*		5.50	6.00	6.00	6.00	6.00	7.00	7.00
Payments to creditors (£000) (1 month's credit)			5.50	6.00	6.00	6.00	6.00	7.00

Information on materials usage costs (£18 per unit) is not relevant because the effect on the cash budget takes place when the bought materials are paid for, not when they are used.

Table 16.2 Phased cash budget for the six months July to December, Year 4

Receipts	*July £000*	*Aug £000*	*Sept £000*	*Oct £000*	*Nov £000*	*Dec £000*
Cash sales	1.50	1.80	2.40	2.40	2.40	3.50
Trade debtors	12.00	14.00	13.50	18.60	19.20	19.80
Dividends				6.00		
Total (R)	13.50	15.80	15.90	27.00	21.60	23.30
Payments						
Trade creditors	5.50	6.00	6.00	6.00	6.00	7.00
Wages	2.60	3.40	3.40	3.40	3.40	3.80
Overheads	1.82	2.38	2.38	2.38	2.38	2.66
Fixed assets – vehicles		15.00				
Taxation			21.00			
Dividends	20.00					
Total (P)	29.92	26.78	32.78	11.78	11.78	13.46
Receipts (R) minus Payments (P)	(16.42)	(10.98)	(16.88)	15.22	9.82	9.84
Opening balance	14.68	(1.74)	(12.72)	(29.60)	(14.38)	(4.56)
Closing balance	(1.74)	(12.72)	(29.60)	(14.38)	(4.56)	5.28

Note: Parentheses indicate overdraft.

It can be seen from the budget shown in Table 16.2 that, on present plans, the company will need an overdraft of over £30 000 in September. As this fact is known sufficiently far in advance of the event, the company can take immediate action to arrange an overdraft facility. If, however, the bank authorises an overdraft but sets a ceiling of, say, £20 000, the company's plans cannot be put into effect. The budget plans would have to be looked at and altered to contain the overdraft requirement within its authorised limit.

Various means can be employed to achieve this objective, either singly or in combination. These are to increase receipts; to reduce payments; to accelerate receipts; and to retard payments. In the present example, the last possibility is the best option. Rephasing the acquisition of vehicles from August to December would solve the problem completely. Alternatively, instead of the vehicles being bought outright and paid for in August, they could be acquired under a leasing contract which would probably require an annual rental of about £4 000. There would thus be a cash saving of £11 000 in August, the difference between the cash price (£15 000) and the leasing rental (£4 000) which would reduce the overdraft requirement to within its limit.

16.3 Construction industry cash budgeting

Before contracts are undertaken it is essential that the cash requirements are budgeted. The procedures are very similar to those operated in the preparation of an ordinary phased cash budget. Preliminary workings are necessary, of the same sort as shown in the final example above. Features found in contract cash requirement assessments are incorporated in the following example.

Example

A contract was budgeted to be completed over a period of six months at a cost of £100 000 and producing a profit of £10 000. Certificates were budgeted to be issued thus:

End of month	*Value (£)*
1	7 000
2	14 000
3	19 000
4	25 000
5	23 000
6	22 000
Total	110 000

Under the terms of the contract, certificates were to be issued during the week following the end of the month to which they related and were to be paid by the client one month after that, subject to a retention of 5 per cent. Of the retention monies, 50 per cent were to be released in Month 7, and 50 per cent after the satisfactory completion of the remedial work in Month 12. Contract costs were budgeted to be paid in the month in which they were incurred.

Required

Prepare a schedule assessing the monthly cash requirements of the contract.

Solution

Month	*Certificates*	*Retentions*	*Receipts*	*Profit*	*Costs paid*
	(1)	*(2) = 5% × (1)*	*(3) = (1)–(2)*	*(4) = 1/11 × (1)*	*(5) = (1)–(4)*
	£	£	£	£	£
1	7 000	350	6 650	636	6 364
2	14 000	700	13 300	1 273	12 727
3	19 000	950	18 050	1 727	17 273
4	25 000	1 250	23 750	2 273	22 727
5	23 000	1 150	21 850	2 091	20 909
6	22 000	1 100	20 900	2 000	20 000
Total	110 000	5 500	104 500	10 000	100 000

These figures are then assembled in a cash budget format, receipts being time-lagged by two months.

Profit is 10/110 or 1/11 of the certificated values and is termed the margin. Cost is arrived at by deducting the profit from the certificated values.

It can be seen from Table 16.3, on the following page, that the cash requirement of the contract would rise from an initial £6 364 in Month 1 to a maximum of £42 000 in Month 5, after which it would reduce and return a cash surplus for the first time in Month 8.

Table 16.3 *Cash requirement assessment schedule (£s)*

	Months										
	1	2	3	4	5	6	7	8	9–11	12	Total
Receipts †			6 650	13 300	18 050	23 750	21 850	20 900	–		110 000
							2 750 *			2 750 *	
Payments	6 364	12 727	17 273	22 727	20 909	20 000					100 000
Monthly requirement	(6 364)	(12 727)	(10 623)	(9 427)	(2 859)	3 750	24 600	20 900	–	2 750	10 000
Brought forward	–	(6 364)	(19 091)	(29 714)	(39 141)	(42 000)	(38 250)	(13 650)	7 250	7 250	
Carried forward	(6 364)	(19 091)	(29 714)	(39 141)	(42 000)	(38 250)	(13 650)	7 250	7 250	10 000	10 000

Notes: † Column (3) time-lagged.
* Release of retentions.

17 Standard Costing

17.1 Introduction

Chapter 15 explained and illustrated the principles of budgetary control. This was seen to be a system of management reporting and control, applied to the business as a whole and to departments within a business and could be described as a macrosystem.

The microversion of this system is termed *standard costing*. Predetermined targets, or standards, are set for individual products, and/or operations, and/or processes. Actual results are recorded and reported in such a way that they can be compared easily with standards. The variances between the two sets of figures are isolated and identified, as a result of which corrective action is taken if the variances are of a significant size.

17.2 Types of standard

Standard costs are the expression in financial terms of the physical targets determined by production personnel. Thus, materials cost standards are the evaluation by the accounting staff of the quantities, types and prices of materials specified by the production department. Similarly, labour cost standards result from the specification of labour hours, grades and rates.

As was noted in Chapter 15, Section 15.2 in connection with budgets, it would be unrealistic to set the targets on the basis of the theoretical maximum level of efficiency which, being an ideal standard, could rarely, if ever, be achieved. Instead, a realistic, attainable level is used as the target which makes due allowance for normal losses and wastage and for normal machine stoppages, termed *downtime*, because of routine maintenance and other factors.

17.3 Preparation of standard costs

The accounting department assembles the physical data obtained from the production engineers, prices them by referring to price lists of materials supplied by the purchasing department and to wages rates obtainable from the personnel or wages departments, and produces the standard cost.

Example

A joinery business manufactures interior wooden doors for supply to building companies. The standard cost data of one type of door favoured by property development companies is:

	Per door
Direct materials	
Wood	1.4 m^2
Direct labour	
Labour	1.6 standard hours

The cost of the wood is ascertained to be £18.00 per square metre, and the wages rate for that labour grade to be £5.50 per hour. From the records of the accounting department, the variable overhead rate per standard hour is calculated to be £6.25 per standard hour.

Using this data it is now a simple matter to calculate the standard cost of manufacturing one door:

Solution

Standard cost of door	£
Direct material	
Wood 1.4 × £18.00	25.20
Direct labour	
Wages 1.6 × £5.50	8.80
Production overhead – variable	
1.6 × £6.25	10.00
Total	44.00

Some accountants would also include an element of fixed production overhead cost in the standard cost, but this involves complications beyond the scope of this book and is not considered further.

When standard costing is used, the standard cost per unit is multiplied by the actual volume of production to give a standard cost of the actual number of units produced. This practice is analogous to flexible budgeting, described in Chapter 15, Section 15.5 and illustrated in Section 15.6.

17.4 Comparison of actual and standard costs

This involves calculating the standard cost of the volume actually produced, as described in Section 17.3, and comparing the corresponding actual figures with it to disclose variances between the two sets of figures.

Example

The facts are the same as in the example given in Section 17.3. An order from a property developer for 2 000 of the doors was completed at the following actual direct costs:

	£
Direct materials	
Wood 3 000 m^2 at £17.00 per m^2	51 000
Direct labour	
Wages 2 600 hours at £7.00 per hour	18 200
Production overheads (variable) totalled £18 000	

On the basis of the actual production of 2 000 doors, the standard cost is calculated as:

Standard cost of 2 000 doors

	£
Direct materials	
Wood 2 000 × £25.20	50 400
Direct labour	
Wages 2 000 × £8.00	17 600
Production overheads variable	
2 000 × £10.00	20 000
Total	88 000

The next step is to compare the two sets of figures and to derive the variances. Variances representing excess of actual costs over standard are labelled 'adverse' (A); variances resulting from actual costs being less than standard are labelled 'favourable' (F).

Actual/standard cost comparison (2 000 doors)

Production	*Standard costs* £	*Actual costs* £	*Variances* £
Direct materials	50 400	51 000	600 (A)
Direct labour	17 600	18 200	600 (A)
Production overhead			
Variable	20 000	18 000	2 000 (F)
Total	88 000	87 200	800 (F)

The variances between the two sets of figures are now the start point for the next stage, that of analysing the variances to identify the factors which have given rise to them.

17.5 Standard cost variance analysis

There are various ways in which variances can be analysed. This section deals exclusively with the traditional approach. Other approaches are too specialised for coverage in this book.

Basically, the analysis of each cost variance assumes that it is attributable to a combination of the degree of utilisation of a physical factor (quantity used or hours taken) and its price (materials list price or wage rate).

This can best be illustrated by means of a diagram in which the actual cost is superimposed on the standard cost, leaving the areas outside the overlap to represent variances. Figure 17.1 shows how this would be done for materials costs. Where, in the diagram, the actual and standard costs coincide (boundaried by the SP and AQ lines and left blank), there are no variances. The variances arise outside this area of overlap and are represented by the shaded rectangles. It is now a simple matter by inspection of the diagram to produce formulae for the calculation of the respective quantity and price variances.

Materials price variance:
$(AQ \times SP) - (AQ \times AP)$

Materials usage variance:
$(SQ \times SP) - (AQ \times SP)$

Other diagrammatical configurations can arise, but the formulae do not alter. Two of the alternative configurations appear in Figure 17.2. The key for Figure 17.2 is identical to that given for Figure 17.1.

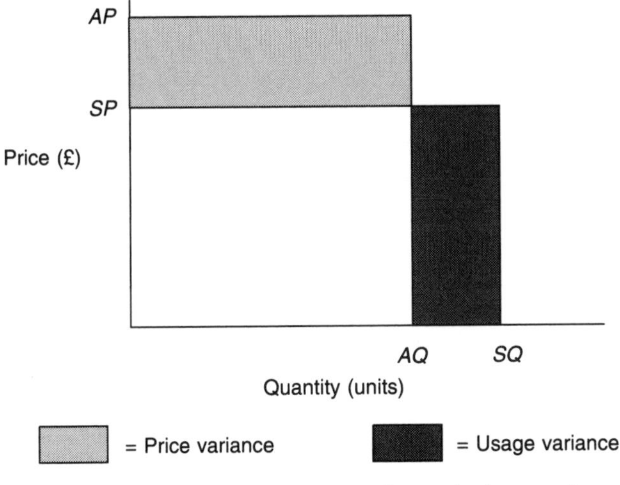

Figure 17.1 Materials standard cost variance

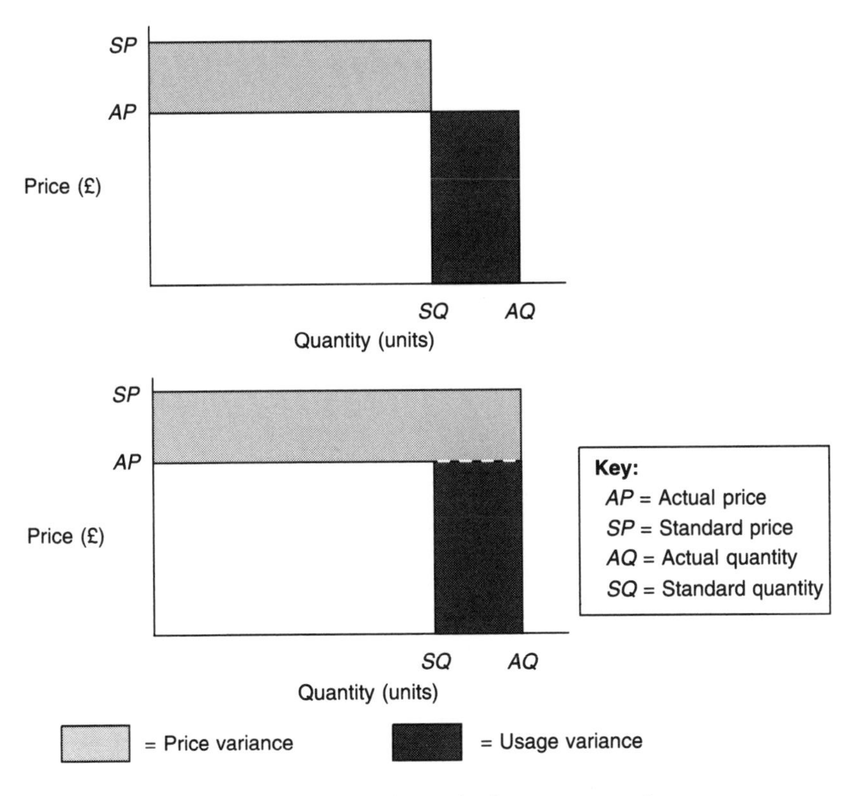

Figure 17.2 Other materials standard cost variance diagrams

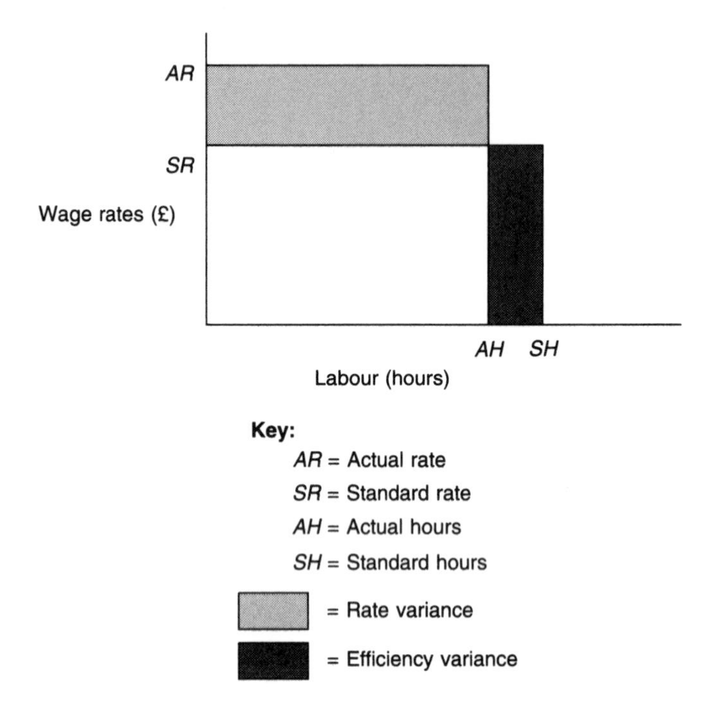

Figure 17.3 Labour standard cost variances

Similar principles apply in the identification of labour cost variances, as shown in Figure 17.3. Formulae can be derived from the above diagram (as they were in the case of materials variances) for the calculation of wages rate and efficiency variances:

Labour rate variance:

$$(AH \times SR) - (AH \times AR)$$

Labour efficiency variance:

$$(SH \times SR) - (AH \times SR)$$

As was illustrated in Figure 17.2, there can be other configurations within the diagram, but these did not affect the formulae for materials variance calculations. The same principle applies also to wages cost variances – the formulae are valid for all alternatives.

The variable overhead standard cost variances follow closely those for wages costs and do not need to be illustrated by a diagram. The formulae are:

Variable overhead expenditure variance:

$(AH \times SR) - (AH \times AR)$

Variable overhead efficiency variance:

$(SH \times SR) - (AH \times SR)$

On the basis of these formulae, standard cost variances can now be analysed.

Example

Use the facts and information given in the examples in Section 17.3 and 17.4 to analyse the standard cost variances

Solution

Figures for the examples cited should be substituted for symbols in the formulae:

	Variance £
Materials price variance	
$(AQ \times SP) - (AQ \times AP)$	
(3 000× £18.00) – (3 000 × £17.00)	
= 54 000 × 51 000	= 3 000 (F)
Materials usage variance	
$(SQ \times SP) - (AQ \times SP)$	
*(2 800 × £18.00) × (3 000 × £18.00)	
= 50 400 × 54 000	= 3 600 (A)
Materials cost variance	600 (A)
*2 000 × 1.4m^2	
Labour rate variance	
$(AH \times SR) - (AH \times AR)$	
(2 600 × £5.50) – (2 600 × £7.00)	
= 14 300 – 18 200	= 3 900 (A)
Labour efficiency variance	
$(SH \times SR) - (AH \times SR)$	
*(3 200 × £5.50) – (2 600 × £5.50)	
= 17 600 – 14 300	= 3 300 (F)
Labour cost variance	600 (A)

*2 000 × 1.6 standard hours

Variable overhead expenditure variance

$(AH \times SR) - (AH \times AR)$

(2 600 × £6.25) − 18 000

= 16 250 − 18 000 = 1750 (A)

Variable overhead efficiency variance

$(SH \times SR) - (AH \times SR)$

(3 200 × £6.25) − (2 600 × £6.25)

= 20 000 − 16 250 = 3 750 (F)

Variable overhead cost variance 2 000 (F)

Each of the total cost variances above is the same as calculated in the comparison schedule in Section 17.4. The analysis of these variances into their constituent elements enables the answers to numerous other questions to be sought. When these have been obtained, action can be taken to eradicate the cause(s) of an adverse variance or to support the underlying reason(s) behind a favourable variance.

To maximise the benefits of a standard cost system, the comparisons have to be made at very frequent, regular intervals. For this reason it is not unusual to find the system working on a weekly, daily or even a shift basis. If the latter is the case, the actual results of a Monday night shift are processed and compared during the Tuesday morning shift and the necessary action taken during the afternoon shift, so that by the start of the Tuesday night shift, the causes giving rise to any significant adverse variances will have been remedied.

17.6 Interrelationship of variances

The fact must not be overlooked that standard cost variances can be interrelated. Taking the variances calculated in Section 17.5 by way of illustration, the following conclusions might be reached. The favourable materials price variance could indicate that cheaper, inferior materials were being used, resulting in more wastage, as shown by the adverse materials usage variance.

With wages, the adverse rate variance might indicate that higher-grade (and therefore, higher-paid) labour was being used than the job required, but that, being more skilled, these workers were able to complete the job in a shorter time, as evidenced by the favourable efficiency variance.

There could also be a link between the wages and materials variances. In an alternative interpretation, the favourable labour efficiency variance might mask the fact that the operatives were rushing the work, resulting in shoddy workmanship and a higher than usual rejection rate and materials wastage, which would account for the adverse materials usage variance.

The foregoing is sufficient to show that when looking at standard cost variances they should never be viewed in isolation, but as part of a much more complex picture.

Appendix I

Extracts from the Annual Report and Accounts 1993 of John Maunders Group plc

(reproduced by permission of the Board of Directors)

JOHN
MAUNDERS
GROUP
plc

Directors' Report

The directors present their annual report on the affairs of the Group together with the accounts and auditors' report for the year ended 30 June 1993.

Principal activities and business review

a. The principal activity of the Group continues to be residential property development.
b. This year the profit before taxation declined to £3,837,000 from £4,227,000 representing a decrease of 9%.
c. Further details of the Group's activities during the year are contained in the Chairman's statement on page 3.

Results and dividends

Group results, dividends and recommended transfers to reserves are as follows:

	£'000	£'000
Group retained profit at beginning of year		16,864
Profit from ordinary activities after taxation and minority interest	2,554	
Dividends per 20p ordinary share		
Interim 2.30p (1992 – 2.30p) paid 30 April 1993	(567)	
Proposed final 2.85p (1992 – 2.65p)	(705)	
		1,282
Group retained profit at end of year		18,146

The final dividend will be paid on 30 November 1993 to Shareholders on the register at close of business on 4 November 1993.

Accounting policies

In the light of recent developments in financial reporting, in particular the publication of the exposure draft 'Reporting the Substance of Transactions in Financial Statements' by the Accounting Standards Board, the directors have reappraised the accounting policies to ensure

JOHN MAUNDERS GROUP plc

compliance with current best practice. As a result, the accounting policies regarding financing of showhouses and recognition of turnover have been revised and prior year figures have been restated.

Directors

The directors of the Company, who are listed on page 10, have held office throughout the year Mr J. B. Davies ceased to be a director of the company on 15 January 1993.

Mr P. D. Kendall and Sir Jeremy Rowe CBE retire by rotation at the next Annual General Meeting and are eligible for re-election.

Mr P. D. Kendall has a service contract subject to six months' written notice.

Directors' interests

The directors who held office at 30 June 1993 had no interests, other than those shown below in the shares of the Company.

	1993			1992		
	20p ordinary shares			20p ordinary shares		
	Trustee	Beneficial	Options	Trustee	Beneficial	Options
J. W. Maunders	1,200,000	5,028,170	–	1,200,000	5,028,170	–
W. H. Bannister	–	10,800	–	–	10,800	
G. R. Swarbrick	1,110,000	10,800	–	1,110,000	10,800	–
P. D. Kendall	–	1,000	30,000	–	1,000	30,000
R. Pinder	–	4,710	30,000	–	–	45,000
Sir Jeremy Rowe CBE	1,000	36,000	–	1,000	36,000	
H. E. Farley	–	10,000	–	–	10,000	–

As at 22 September 1993 the interests of the directors in the shares of the Company were as stated above.

Included in the non-beneficial holding of Mr J. W. Maunders are 900,000 20p ordinary shares which are held on trust for his children and which are duplicated in the non-beneficial holding of Mr G. R. Swarbrick.

During the year options were exercised in respect of 15,000 20p ordinary shares by Mr R. Pinder. None of the directors had any interest in any material contract with the Company during the year.

Share Option Scheme (1985)

The Company has granted, for no consideration, options to subscribe for 584,000 ordinary shares. These options are exercisable between three and ten years from the date of grant.

Date of Grant	Number of shares	Subscription price per share
2 November 1987	107,000	89.33p
17 October 1988	57,000	103.60p
10 April 1989	36,000	112.60p
16 October 1989	30,000	79.00p
8 April 1991	168,000	120.40p
11 November 1991	54,000	135.00p
6 April 1992	66,000	135.60p
3 May 1993	66,000	141.40p

**JOHN
MAUNDERS
GROUP
plc**

Sustantial Shareholders

At 22 September 1993 the following interests of more than 3% of the Company's issued share capital had been notified to the Company:

	Number of shares	% of share capital
Ms W. R. Maunders	2,651,830	10.72
Scottish Amicable		
Investment Managers Limited	1,505,000	6.08
Electricity Supply Pension Scheme	1,195,000	4.83
J. W. Maunders Settlement	900,000	3.34
State Street Bank & Trust Company	831,100	3.36
Henderson Administration Group PLC	792,000	3.20
Friends Provident Group	750,000	3.03

Corporate Governance

The directors have studied the Code of Best Practice published in December 1992 by the Committee on the Financial Aspects of Corporate Governance. The Company already complies with a number of the Committee's recommendations and others including the formation of an audit committee and a remuneration committee, are about to be implemented.

The remaining issues will be considered in the coming year in the light of the anticipated further guidance and having regard to practicality.

The non-executive directors are confident that there is a free flow of information and discussion between Board members, that all directors are kept informed in a timely manner and are in a position to continue to contribute towards the success of the Group.

Close company status

The Company is not a close company within the provisions of the Income and Corporation Taxes Act 1988.

Fixed assets

Information relating to changes in tangible fixed assets is given in note 11 to the accounts.

Charitable and political contributions

The Group contributed £6,987, to charities. There were no contributions to political parties.

Employees

Special attention is given to health and safety. It is Group policy to support the employment of disabled persons wherever practicable and to ensure wherever possible they share in the training, career development and promotion opportunities available to all employees. The Group keeps employees informed by direct contact, internal notification and also through publicly available sources such as press releases, annual accounts and circulars to Shareholders.

JOHN
MAUNDERS
GROUP
plc

Auditors

Messrs. Arthur Andersen resigned during the year, confirming there were no circumstances they considered should be brought to the notice of the members or the creditors of the Company, and Price Waterhouse were appointed to fill the vacancy. Special notice having been received a resolution will be placed before the Annual General Meeting, as special business, to reappoint Price Waterhouse as auditors.

Development House
Crofts Bank Road
Urmston
Manchester M31 1UH

By order of the Board
G. R. Swarbrick
Secretary

22 October 1993

Auditors' Report

To the Members of John Maunders Group plc

We have audited the accounts on pages 16 to 29 in accordance with Auditing Standards.

In our opinion, the accounts give a true and fair view of the state of affairs of the Company and the Group at 30 June 1993 and of the profit and cash flows of the Group for the year then ended and have been properly prepared in accordance with the Companies Act 1985.

Price Waterhouse
Chartered Accountants and Registered Auditor
York House
York Street
Manchester M2 4WS

22 October 1993

JOHN MAUNDERS GROUP plc

Statement of Accounting Policies

In the light of recent developments in financial reporting, in particular the publication of the exposure draft 'Reporting the Substance of Transactions in Financial Statements' by the Accounting Standards Board, the directors have reappraised the accounting policies to ensure compliance with current best practice. As a result, the accounting policies regarding (i) financing of showhouses and (j) recognition of turnover have been revised and prior year figures have been restated.

A summary of the principal Group accounting policies is set out below.

(a) **Basis of accounting**
The accounts have been prepared under the historical cost convention, modified to include the revaluation of investment properties. The Group accounts have been prepared in accordance with applicable accounting standards.

(b) **Basis of consolidation**
The Group accounts consolidate the accounts of John Maunders Group plc and its subsidiary undertakings made up to 30 June 1993, after eliminating all significant inter company balances and transactions.

No profit and loss account is presented for John Maunders Group plc as provided by s.230 of the Companies Act 1985.

(c) **Intangible fixed assets**
Ground rents created on the sale of properties are capitalised at a valuation which represents seven years' rent receivable.

(d) **Tangible fixed assets**
Freehold land and buildings, computer equipment and software, plant, equipment and motor vehicles are stated at cost, less accumulated depreciation. Depreciation is provided at rates calculated to write-off the cost, less estimated residual value, of each asset over its expected useful life as follows:

Freehold buildings	2% p.a. (straight line)
Computer equipment and software	33⅓% p.a. (straight line)
Plant and equipment	10–25% p.a. (reducing balance)
Motor vehicles	33⅓% p.a. (reducing balance)

Profits or losses on the disposal of fixed assets are included in the calculation of operating profit.

Investment properties are revalued annually and the aggregate surplus or deficit is transferred to revaluation reserve, and no provision is made for depreciation. This departure from the requirements of the Companies Act 1985, which requires all properties to be depreciated, is, in the opinion of the directors, necessary for the accounts to show a true and fair view in accordance with applicable accounting standards. The amounts involved are not material.

(e) **Fixed asset investments**
In the Company's accounts investments in subsidiary undertakings are stated at cost, less amounts written-off plus the Group's share of post-acquisition retained profits and reserves, with a corresponding credit to the revaluation reserve. The directors consider that this policy fairly represents the value of the Company's investment.

JOHN MAUNDERS GROUP plc

(f) **Land for development**
Land for development is stated at the lower of cost and net realisable value.

(g) **Stocks and work-in-progress**
Stocks and work-in-progress are stated at the lower of cost and net realisable value.

Building materials	–	purchase cost on a first-in, first-out basis
Work-in-progress	–	cost of direct materials and labour, plus an appropriate proportion of building overheads based on normal levels of activity
Houses for resale	–	lower of purchase price plus incidental costs and net realisable value

Net realisable value is based on estimated normal selling price less further costs expected to be incurred in completion and disposal.

(h) **Taxation**
Corporation tax payable is provided on taxable profits at the current rate.
Provision for deferred taxation is made under the liability method only to the extent that tax timing differences are expected to reverse in the future without being replaced, calculated at the rate at which it is anticipated the timing differences will reverse.

(i) **Financing of showhouses**
The Group has revised its accounting policy for the financing of showhouses and the recognition of profits and losses arising. Showhouses were previously held off balance sheet under a licence agreement. The accounts now reflect showhouses within land for development and stock and work-in-progress (at the lower of cost and net realisable value), and the related borrowings as creditors. These borrowings are secured on the freehold title of the showhouses. The profits and losses realised on the sale of these properties, previously taken at the commencement of the licence transaction, are now recognised on the eventual sale.

(j) **Turnover**
The Group has revised its policy for the recognition of turnover and profits in relation to new residential properties. Turnover and profits were previously recognised when sales of new residential properties were financially complete. Turnover and profits are now recognised on legal completion. Resales of second-hand properties sold under the Group's part exchange scheme are not included in turnover. All sales are made in the United Kingdom.

(k) **Pension costs**
It is the policy of the Group to fund pension liabilities, on the advice of external actuaries, by payments to independent trusts.
Independent actuarial valuations are carried out every three years. Any resulting surpluses or deficits are credited or charged to the profit and loss account as variations from the regular cost over the average remaining service lives of members.

(l) **Hire purchase contracts**
Assets held under hire purchase contracts are capitalised and depreciated over their expected useful lives. Outstanding commitments are included in creditors and finance charges are written-off over the period of the contract.

JOHN MAUNDERS GROUP plc

Consolidated Profit and Loss Account

For the year ended 30 June 1993

	Notes	**1993**			**1992**
					Restated
		£'000	**£'000**	£'000	£'000
Turnover	1		**59,809**		52,023
Cost of sales			**(50,774)**		(42,444)
Gross profit			**9,035**		9,579
Operating expenses, net	2		**(3,740)**		(3,596)
Operating profit			**5,295**		5,983
Investment income	3		**78**		77
Interest payable and similar charges	4		**(1,536)**		(1,833)
Profit from ordinary activities before taxation	**5**		**3,837**		4,227
Taxation	7		**(1,284)**		(1,395)
Profit from ordinary activities after taxation			**2,553**		2,832
Minority interest			**1**		–
Profit for the financial year			**2,554**		2,832
Dividends	8		**(1,272)**		(1,219)
Retained profit for the year, of which £1,282,000 (1992 – £1,613,000) is dealt with in the accounts of the holding company			**1,282**		1,613
Retained profit brought forward					
As previously reported		**17,887**		16,423	
Prior year adjustment		**(1,023)**		(1,172)	
As restated			**16,864**		15,251
Retained profit carried forward			**18,146**		16,864
Earnings per ordinary share	9		**10.36 p**		11.52 p

The accompanying notes are an integral part of this consolidated profit and loss account.

JOHN MAUNDERS GROUP plc

Balance Sheets at 30 June

	Notes	Group 1993 £'000	Group 1992 Restated £'000	Company 1993 £'000	Company 1992 Restated £'000
Fixed assets					
Intangible assets	10	**425**	398	**425**	398
Tangible assets	11	**1,726**	1,710	**1,726**	1,710
Investments	12	**2**	2	**332**	331
		2,153	2,110	**2,483**	2,439
Current assets					
Land for development		**26,642**	31,466	**26,642**	31,466
Stocks and work-in-progress	13	**13,612**	13,531	**13,612**	13,531
Debtors	14	**1,974**	1,475	**1,991**	1,492
Cash at bank and in hand		**557**	23	**557**	9
		42,785	46,495	**42,802**	46,498
Creditors: Amounts falling due within one year	15	**18,525**	23,094	**18,872**	23,432
Net current assets		**24,260**	23,401	**23,930**	23,066
Total assets less current liabilities		**26,413**	25,511	**26,413**	25,505
Creditors: Amounts falling due after more than one year	16	**413**	892	**413**	892
Provisions for liabilities and charges	17	**93**	84	**93**	84
Net assets		**25,907**	24,535	**25,907**	24,529
Capital and reserves					
Called up share capital	18	**4,936**	4,924	**4,936**	4,924
Share premium	19	**2,624**	2,590	**2,624**	2,590
Revaluation reserve	19	**201**	151	**295**	245
Profit and loss account	19	**18,146**	16,864	**18,052**	16,770
Shareholders' funds		**25,907**	24,529	**25,907**	24,529
Minority interest		**–**	6	**–**	–
Total capital employed		**25,907**	24,535	**25,907**	24,529

Signed on behalf of the Board

J.W. Maunders }
G. R. Swarbrick } Directors

22 October 1993

The accompanying notes are an integral part of these balance sheets.

JOHN
MAUNDERS
GROUP
plc

Consolidated cash flow statement

For the year ended 30 June 1993

	Notes	1993 £'000	1992 Restated £'000
Net cash inflow from operating activities	20	**12,018**	4,656
Returns on investments and servicing of finance			
Investment income received		**81**	95
Interest received		–	–
Interest paid		**(1,537)**	(1,852)
Interest element of hire purchase payments		**(28)**	(28)
Dividends paid		**(1,220)**	(1,213)
Net cash outflow from returns on investments and servicing of finance		**(2,704)**	(2,997)
Corporation tax paid		**(1,364)**	(1,262)
Investing activities			
Payments to acquire fixed assets		**(209)**	(177)
Receipts from sales of tangible fixed assets		**56**	51
Net cash outflow from investing activities		**(153)**	(126)
Net cash inflow before financing		**7,797**	271
Financing			
Capital element of hire purchase payments	21	**(237)**	(242)
Decrease in secured loan	21	**(427)**	(876)
Issue of ordinary share capital	21	**46**	185
Net cash outflow from financing		**(618)**	(933)
Increase (decrease) in cash and cash equivalents	22	**7,179**	(662)

The accompanying notes are an integral part of this consolidated cash flow statement.

JOHN MAUNDERS GROUP plc

Consolidated statement of recognised gains and losses

For the year ended 30 June 1993

	1993	1992 Restated
	£'000	£'000
Profit for the financial year	**2,554**	**2,832**
Unrealised surplus on revaluation of properties	**50**	–
Total recognised gains relating to the year	**2,604**	2,832
Prior year adjustment	**(1,023)**	
Total recognised gains since last annual report	**1,581**	

Reconciliation of movements in shareholders' funds

For the year ended 30 June 1993

		1993	1992 Restated
		£'000	£'000
Profit for the financial year		**2,554**	2,832
	Dividends	**(1,272)**	(1,219)
		1,282	1,613
Unrealised surplus on revaluation of properties		**50**	–
New share capital subscribed		**46**	185
Net increase in shareholders' funds		**1,378**	1,798
Opening shareholders' funds		**24,529**	22,731
Closing shareholders' funds		**25,907**	24,529

The accompanying notes are an integral part of these consolidated statements.

JOHN MAUNDERS GROUP plc

Notes to the accounts

For the year ended 30 June 1993

1. Segment information

All of the Group turnover, profit from ordinary activities before taxation and net assets are derived from residential property development within the United Kingdom.

2. Operating expenses

	1993	1992
	£'000	£'000
Administrative expenses	**3,831**	3,659
Other operating income	**(91)**	(63)
	3,740	3,596

3. Investment income

	1993	1992
	£'000	£'000
Ground rent income	**58**	56
Rental income from investment properties	**20**	20
Interest receivable	–	1
	78	77

4. Interest payable and similar charges

	1993	1992
	£'000	£'000
On bank loans, overdrafts and other loans repayable within five years		
– by instalment	**36**	40
– not by instalment	**1,153**	1,349
	1,189	1,389
On secured loans	**347**	444
	1,536	1,833

5. Profit from ordinary activities before taxation

	1993	1992
	£'000	£'000
Profit from ordinary activities before taxation is stated after charging (crediting):		
Loss (profit) on sale of tangible fixed assets	**4**	(1)
Depreciation	**305**	332
Hire of plant and machinery	**325**	311
Auditors' remuneration	**25**	30
Staff costs (note 6)	**3,634**	3,493
Operating lease (recoveries) rentals	**(23)**	65

JOHN MAUNDERS GROUP plc

6. Staff costs

	1993	1992
	£'000	£'000
Employee costs, including directors, comprised:		
Wages and salaries	**3,137**	3,002
Social security costs	**264**	275
Other pension costs	**233**	216
	3,634	3,493

The average weekly number of persons directly employed by the Group during the year was as follows:	**1993**	1992
	Number	Number
Building	**36**	33
Technical	**20**	23
Sales and marketing	**49**	44
Administration	**30**	34
Legal	**5**	5
	140	139

Directors' remuneration

The employee costs, shown above, include the following remuneration in respect of directors of John Maunders Group plc:	**1993**	1992
	£'000	£'000
Fees	**30**	26
Emoluments, including pension contributions	**589**	608
Performance related payments	**88**	155
	707	789

Emoluments excluding pension contributions:	**1993**	1992
Chairman and highest paid director	**£'000**	£'000
Emoluments	**162**	161
Performance related payments	**36**	48
	198	209

The directors received emoluments, excluding pension contributions, in the following ranges:	**1993**	1992
	Number	Number
£10,001 – £15,000	**–**	1
£15,001 – £20,000	**2**	1
£60.001 – £65,000	**1**	–
£70,001 – £75,000	**1**	1
£85,001 – £90,000	**–**	1
£90,001 – £95,000	**2**	–
£95,001 – £100,000	**–**	1
£100,001 – £105,000	**–**	1

JOHN MAUNDERS GROUP plc

The directors received emoluments, excluding pension contributions, in the following ranges *continued*:	**1993** Number	1992 Number
£105,001 – £110,000	**1**	–
£140,001 – £145,000	**–**	1
£195,001 – £200,000	**1**	–
£205,001 – £210,000	**–**	1

7. Taxation

	1993	1992
	£'000	£'000
The Group taxation charge comprises:		
UK corporation tax on the profit for the year at 33%	**1,068**	1,395
Deferred tax	**216**	
	1,284	1,395

8. Dividends

	1993	1992
	£'000	£'000
Interim paid of 2.30p per share (1992 – 2.30p per share)	**567**	566
Final proposed of 2.85p per share (1992 – 2.65p per share)	**705**	653
	1,272	1,219

9. Earnings per ordinary share

Earnings per ordinary share are based on Group profit from ordinary activities after taxation and minority interest and on 24,657,301 (1992 – 24,584,171) ordinary shares being the weighted average number of ordinary shares in issue during the year adjusted for the exercise of options for 57,000 ordinary shares.

10. Intangible fixed assets

	Group and Company	
	1993	1992
Intangible fixed assets comprise ground rents.	**£'000**	£'000
The movement during the year was as follows:		
Balance at beginning of year	**398**	335
Additions	**27**	64
Disposals	**–**	(1)
Balance at end of year	**425**	398

JOHN MAUNDERS GROUP plc

11. Tangible fixed assets

The movements in the year for the Group and Company were as follows:

	Investment properties	Freehold land and buildings	Plant, equipment and motor vehicles	Total
Cost or valuation	**£'000**	**£'000**	**£'000**	**£'000**
Beginning of year	200	984	1,919	3,103
Revaluation surplus	55	–	–	55
Additions	–	–	326	326
Disposals	–	–	(186)	(186)
End of year	255	984	2,059	3,298
Depreciation				
Beginning of year	–	153	1,240	1,393
Charge	–	17	288	305
Disposals	–	–	(126)	(126)
	–	170	1,402	1,572
Net book value, end of year	255	814	657	1,726

Investment properties were valued on 30 June 1993 on an open market existing use basis by external chartered surveyors.

Included in freehold land and buildings is land at a cost of £107,000 which is not depreciated. The net book value of plant, equipment and motor vehicles includes £133,000 (1992 – £276,800) in respect of assets held under hire purchase contracts.

12. Fixed asset investments

Fixed asset investments consist of interests in subsidiary undertakings, all of which are incorporated and operate in England, and other sundry investments. Interests in subsidiary undertakings are as follows:

	Interest	Activity
Delany Brothers (House Builders) Limited	100%	Non-trading
Haven Retirement Homes Limited	100%	Non-trading
J.W. Liptrot & Company Limited	100%	Non-trading
Maunders Homes (East Anglia) Limited	100%	Non-trading
Maunders Homes (North West) Limited	100%	Non-trading
Maunders Homes (South) Limited	100%	Non-trading
Maunders Urban Renewal Limited	100%	Non-trading
Maunders Inner City Limited	100%	Non-trading

JOHN MAUNDERS GROUP plc

12. Fixed asset investments *continued*

Maunders Homes (East Anglia) Limited, Maunders Homes (North West) Limited, Maunders Homes (South) Limited, Maunders Urban Renewal Limited and Maunders Inner City Limited are sales agencies for the Group.

At an extraordinary meeting of Mancunian Developments Limited held on 2 July 1992, it was resolved that the company be wound-up voluntarily. The return of final meeting was filed with the Registrar of Companies on 24 February 1993.

Fixed asset investments:	Group		Company	
	1993	1992	**1993**	1992
	£'000	£'000	**£'000**	£'000
Subsidiary undertakings	**–**	–	**330**	329
Other investments (listed in the UK)	**2**	2	**2**	2
	2	2	**332**	331

Investments in subsidiary undertakings:	**1993**	1992
	£'000	£'000
Balance at beginning of year	**329**	329
Share of current year profits	**1**	–
Balance at end of year	**330**	329

13. Stocks and work-in-progress

	Group & Company	
	1993	1992
	£'000	£'000
Building materials and showhouse contents	**887**	759
Work-in-progress	**9,134**	10,548
Houses held for resale	**3,591**	2,224
	13,612	13,531

14. Debtors

	Group		Company	
Amounts falling due within one year	**1993**	1992	**1993**	1992
	£'000	£'000	**£'000**	£'000
Trade debtors	**964**	291	**964**	291
Due from subsidiary undertakings	**–**	–	**17**	17
Other debtors	**458**	343	**458**	343
Prepayments	**197**	189	**197**	189
Deferred tax recoverable	**–**	216	**–**	216
	1,619	1,039	**1,636**	1,056
Amounts falling due after one year				
Trade debtors	**150**	236	**150**	236
ACT recoverable	**205**	200	**205**	200
	1,974	1,475	**1,991**	1,492

JOHN MAUNDERS GROUP plc

15. Creditors

Amounts falling due within one year	Group		Company	
	1993	1992	**1993**	1992
	£'000	£'000	**£'000**	£'000
Bank overdrafts	**3,437**	10,082	**3,437**	10,082
Secured loan	**3,050**	3,024	**3,050**	3,024
Trade creditors	**4,328**	3,163	**4,328**	3,163
Land creditors	**3,727**	2,801	**3,727**	2,801
Due to subsidiary undertakings	**–**	–	**343**	334
UK corporation tax	**1,073**	1,360	**1,077**	1,364
Taxation and social security	**178**	199	**178**	199
Amounts due under hire purchase contracts	**80**	174	**80**	174
Proposed dividend	**705**	653	**705**	653
Accruals	**1,947**	1,638	**1,947**	1,638
	18,525	23,094	**18,872**	23,432

Bank overdrafts are secured by unlimited cross guarantees between all group undertakings and by specific legal charges on land and work-in-progress. The freehold titles of the associated showhouses have been pledged as security for the secured loan

16. Creditors

Amounts falling due after more than one year	Group and Company	
	1993	1992
	£'000	£'000
Secured loan	**3,462**	3,889
Hire purchase creditors	**81**	201
Included in creditors: amounts falling due within one year (note 15)	**(3,130)**	(3,198)
	413	892

Long-term borrowings are repayable as follows:	Group and Company	
	1993	1992
	£'000	£'000
1 – 2 years	**413**	891
2 – 5 years	**–**	1
	413	892

JOHN MAUNDERS GROUP plc

17. Provisions for liabilities and charges

	Group and Company	
	1993	1992
Provisions for liabilities and charges comprise:	**£'000**	£'000
Deferred taxation	**14**	14
Warranties	**79**	70
	93	84

	Group and Company	
	1993	1992
The provision for deferred taxation is attributable to:	**£'000**	£'000
Excess of tax allowances over book depreciation of fixed assets	**–**	22
ACT recoverable	**–**	(17)
	–	5
Capital gain arising on revaluation of investment properties	**14**	9
	14	14

18. Called up share capital

	1993	1992
	£'000	£'000
Authorised		
30,000,000 (1992 – 30,000,000) ordinary shares of 20p each	**6,000**	6,000
Allotted, called-up and fully paid		
24,677,500 (1992 – 24,620,500) ordinary shares of 20p each	**4,936**	4,924

During the year the Group allotted 57,000 20p ordinary shares for cash with a nominal value of £11,400 and at a premium of £34,210 in respect of options exercised under the Share Option Scheme (1985).

19. Reserves

		Group	
	Share Premium	Revaluation Reserve	Profit and Loss Account
	£'000	£'000	£'000
Balance at beginning of year	2,590	151	16,864
Shares issued	34	–	–
Surplus arising on revaluation	–	50	–
Retained profit for the year	–	–	1,282
Balance at end of year	2,624	201	18,146

JOHN MAUNDERS GROUP plc

The Group profit and loss account at 30 June 1993 includes £94,000 (1992 – £94,000) relating to post acquisition profits of subsidiary undertakings. In the Company balance sheet, these profits (which are non-distributable) are included in the revaluation reserve.

20. Net cash inflow from operating activities

	1993	1992
	£'000	£'000
Reconciliation of operating profit to net cash inflow from operating activities		
Operating profit	**5,295**	5,983
Depreciation	**305**	332
Minority interest	**(5)**	–
Ground rents capitalised	**(27)**	(64)
Loss (profit) on sale of tangible fixed assets	**4**	(1)
Decrease in land for development	**4,824**	1,180
Increase in stock and work-in-progress	**(81)**	(24)
Increase in debtors	**(701)**	(190)
Increase (decrease) in creditors and provisions	**2,404**	(2,560)
	12,018	4,656

21. Financing

	Share Capital and Share Premium £'000	Secured Loan £'000	Hire Purchase Obligations £'000
Analysis of changes in financing during the year			
Balance at beginning of year	7,514	3,889	201
New hire purchase agreements	–	–	117
Net cash inflow (outflow) from financing	46	(427)	(237)
Balance at end of year	7,560	3,462	81

22. Increase (Decrease) in cash and cash equivalents

	1993	1992
	£'000	£'000
Analysis of changes in cash and cash equivalents during the year		
Balance at beginning of year	**(10,059)**	(9,397)
Net cash inflow (outflow)	**7,179**	(662)
Balance at end of year	**(2,880)**	(10,059)

**JOHN
MAUNDERS
GROUP
plc**

Increase (Decrease) in cash and cash equivalents *continued*

Analysis of the balances of cash and cash equivalents as shown in the balance sheet	**1993 £'000**	1992 £'000	Change in year £'000
Cash at bank and in hand	**557**	23	534
Bank overdraft	**(3,437)**	(10,082)	6,645
	(2,880)	(10,059)	7,179

23. Financial commitments

	1993 £'000	1992 £'000
(a) The Group's capital commitments are as follows:		
Contracted	**374**	239
Authorised but not contracted	–	–
	374	239

(b) The Group has entered into options and conditional contracts, to a value of £9,010,000, to purchase land for development.

24. Contingent liabilities

The Group has, in the normal course of business, given counter indemnities in connection with performance and other guarantees.

25. Pension arrangements

The Group maintains a contributory pension scheme covering substantially all its employees. The scheme, which is funded by payments to independent trusts, provides defined benefits based on the length of service and pensionable salary at retirement. It is the policy of the Group to fund the liabilities of the scheme over the lifetime of the existing members. The funding policy and the accounting policy are the same. The pension costs of the scheme to the Group were £233,000 (1992 – £216,000).

An actuarial valuation of the scheme was prepared as at 1 September 1991 by Scottish Widows' Fund and Life Assurance Society. The contribution rate was assessed using the projected unit method and the valuation assumed a rate of interest of 1.5% p.a. in excess of the rate of salary increases and no discretionary increases in benefits.

The 1991 actuarial valuation of the scheme showed that (i) the existing assets fully cover liabilities based on current salaries, (ii) the estimated market value of the assets was £2,117,500 and (iii) the funding level at that date, taking into account future salary increases, was 123%.

The Group, after consultation with the actuaries, has continued to fund the scheme at the existing level.

JOHN MAUNDERS GROUP plc

Five-year summary

		1993	1992 (Restated)	1991 (Restated)	1990 (Restated)	1989 (Restated)
		£'000	£'000	£'000	£'000	£'000
Profit and loss account						
Turnover		**59,809**	52,022	52,189	57,667	50.674
Operating profit		**5,295**	5,983	5,795	8,871	8.500
Profit before taxation		**3,837**	4,227	3,277	5,830	6,126
Profit after taxation and minority interest		**2,553**	2,832	2,158	3,794	3,959
Balance sheet						
Shareholders' funds		**25,907**	24.529	22,924	21,858	19,266
Tangible fixed assets		**1,726**	1,710	1,616	1,609	1,732
Net current assets		**24,260**	23,401	24,306	24,801	18,237
Statistical information						
Operating profit on turnover	%	**8.85**	11.50	11.10	15.38	16.77
Profit before tax and minority interest on turnover	%	**6.42**	8.13	6.28	10.11	12.09
Profit before tax and minority interest on shareholders' funds	%	**14.81**	17.23	14.30	26.67	31.80
Earnings per ordinary share p		**10.4**	11.5	8.8	15.5	16.2
Dividend per ordinary share	p	**5.15**	4.95	4.95	4.95	4.95
Shareholders' funds per ordinary share	p	**104.98**	99.63	93.87	89.98	79.31
Current asset ratio		**2.31**	2.01	2.06	2.01	1.67
Number of legal completions		**904**	775	664	696	654

Notes:

(a) The minority interest arose on the formation of Mancunian Developments Limited in November 1984.

(b) Earnings, dividend and shareholders' funds per ordinary share have been adjusted to take account of capitalisation issues, rights issues and the exercise of share options.

Appendix II

Present Value Tables

Percentage rate of discount

Future years	1	2	3	4	5	6	7	8	9	10	11	12	13	14	15	16
1	0.990	0.980	0.971	0.962	0.952	0.943	0.935	0.926	0.917	0.909	0.901	0.893	0.885	0.877	0.870	0.862
2	0.980	0.961	0.943	0.925	0.907	0.890	0.873	0.857	0.842	0.826	0.812	0.797	0.783	0.769	0.756	0.743
3	0.971	0.942	0.915	0.889	0.864	0.840	0.816	0.794	0.772	0.751	0.731	0.712	0.693	0.675	0.658	0.641
4	0.961	0.924	0.888	0.855	0.823	0.792	0.763	0.735	0.708	0.683	0.659	0.636	0.613	0.592	0.572	0.552
5	0.951	0.906	0.863	0.822	0.784	0.747	0.713	0.681	0.650	0.621	0.593	0.567	0.543	0.519	0.497	0.476
6	0.942	0.888	0.837	0.790	0.746	0.705	0.666	0.630	0.596	0.564	0.535	0.507	0.480	0.456	0.432	0.410
7	0.933	0.871	0.813	0.760	0.711	0.665	0.623	0.583	0.547	0.513	0.482	0.452	0.425	0.400	0.376	0.354
8	0.923	0.853	0.789	0.731	0.677	0.627	0.582	0.540	0.502	0.467	0.434	0.404	0.376	0.351	0.327	0.305
9	0.914	0.837	0.766	0.703	0.645	0.592	0.544	0.500	0.460	0.424	0.391	0.361	0.333	0.308	0.284	0.263
10	0.905	0.820	0.744	0.676	0.614	0.558	0.508	0.463	0.422	0.386	0.352	0.322	0.295	0.270	0.247	0.227
11	0.896	0.804	0.722	0.650	0.585	0.527	0.475	0.429	0.388	0.350	0.317	0.287	0.261	0.237	0.215	0.195
12	0.887	0.788	0.701	0.625	0.557	0.497	0.444	0.397	0.356	0.319	0.286	0.257	0.231	0.208	0.187	0.168
13	0.879	0.773	0.681	0.601	0.530	0.469	0.415	0.368	0.326	0.290	0.258	0.229	0.204	0.182	0.163	0.145
14	0.870	0.758	0.661	0.577	0.505	0.442	0.388	0.340	0.299	0.263	0.232	0.205	0.181	0.160	0.141	0.125
15	0.861	0.743	0.642	0.555	0.481	0.417	0.362	0.315	0.275	0.239	0.209	0.183	0.160	0.140	0.123	0.108
16	0.853	0.728	0.623	0.534	0.458	0.394	0.339	0.292	0.252	0.218	0.188	0.163	0.141	0.123	0.107	0.093
17	0.844	0.714	0.605	0.513	0.436	0.371	0.317	0.270	0.231	0.198	0.170	0.146	0.125	0.108	0.093	0.080
18	0.836	0.700	0.587	0.494	0.416	0.350	0.296	0.250	0.212	0.180	0.153	0.130	0.111	0.095	0.081	0.069
19	0.828	0.686	0.570	0.475	0.396	0.331	0.277	0.232	0.194	0.164	0.138	0.116	0.098	0.083	0.070	0.060
20	0.820	0.673	0.554	0.456	0.377	0.312	0.258	0.215	0.178	0.149	0.124	0.104	0.087	0.073	0.061	0.051

Percentage rate of discount

Future years	17	18	19	21	21	22	23	24	25	26	28	30	35	40	45	50
1	0.855	0.847	0.840	0.833	0.826	0.820	0.813	0.806	0.800	0.794	0.781	0.769	0.741	0.714	0.690	0.667
2	0.731	0.718	0.706	0.694	0.683	0.672	0.661	0.650	0.640	0.630	0.610	0.592	0.549	0.510	0.476	0.444
3	0.624	0.609	0.593	0.579	0.564	0.551	0.537	0.524	0.512	0.500	0.477	0.455	0.406	0.364	0.328	0.296
4	0.534	0.516	0.499	0.482	0.467	0.451	0.437	0.423	0.410	0.397	0.373	0.350	0.301	0.260	0.226	0.198
5	0.456	0.437	0.419	0.402	0.386	0.370	0.355	0.341	0.328	0.315	0.291	0.269	0.223	0.186	0.156	0.132
6	0.390	0.370	0.352	0.335	0.319	0.303	0.289	0.275	0.262	0.250	0.227	0.207	0.165	0.133	0.108	0.088
7	0.333	0.314	0.296	0.279	0.263	0.249	0.235	0.222	0.210	0.198	0.178	0.159	0.122	0.095	0.074	0.059
8	0.285	0.266	0.249	0.233	0.218	0.204	0.191	0.179	0.168	0.157	0.139	0.123	0.091	0.068	0.051	0.039
9	0.243	0.225	0.209	0.194	0.180	0.167	0.155	0.144	0.134	0.125	0.108	0.094	0.067	0.048	0.035	0.026
10	0.208	0.191	0.176	0.162	0.149	0.137	0.126	0.116	0.107	0.099	0.085	0.073	0.050	0.035	0.024	0.017
11	0.178	0.162	0.148	0.135	0.123	0.112	0.103	0.094	0.086	0.079	0.066	0.056	0.037	0.025	0.017	0.012
12	0.152	0.137	0.124	0.112	0.102	0.092	0.083	0.076	0.069	0.062	0.052	0.043	0.027	0.018	0.012	0.008
13	0.130	0.116	0.104	0.093	0.084	0.075	0.068	0.061	0.055	0.050	0.040	0.033	0.020	0.013	0.008	0.005
14	0.111	0.099	0.088	0.078	0.069	0.062	0.055	0.049	0.044	0.039	0.032	0.025	0.015	0.009	0.006	0.003
15	0.095	0.084	0.074	0.065	0.057	0.051	0.045	0.040	0.035	0.031	0.025	0.020	0.011	0.006	0.004	0.002
16	0.081	0.071	0.062	0.054	0.047	0.042	0.036	0.032	0.028	0.025	0.019	0.015	0.008	0.005	0.003	0.002
17	0.069	0.060	0.052	0.045	0.039	0.034	0.030	0.026	0.023	0.020	0.015	0.012	0.006	0.003	0.002	0.001
18	0.059	0.051	0.044	0.038	0.032	0.028	0.024	0.021	0.018	0.016	0.012	0.009	0.005	0.002	0.001	0.001
19	0.051	0.043	0.037	0.031	0.027	0.023	0.020	0.017	0.014	0.012	0.009	0.007	0.003	0.002	0.001	0.000
20	0.043	0.037	0.031	0.026	0.022	0.019	0.016	0.014	0.012	0.010	0.007	0.005	0.002	0.001	0.001	0.000

Appendix III

CIOB Member Examination Part II: Building Economics and Finance, Questions (extracts) and Suggested Solutions

(The questions are reproduced by permission of the Chartered Institute of Building.)

The questions in this Appendix have been selected from the Building Economics and Finance examination papers (Member examination Part II) of the Chartered Institute of Building.

Suggested solutions, which are entirely the author's own and begin on page 217, are cross-referenced to chapter and section numbers within the main body of the book.

In order to obtain the maximum benefit from the questions, you should attempt to answer them initially without reference to the solutions. These can subsequently be consulted to confirm your own answer and to act as a safety net if you find the question too difficult.

QUESTIONS

Chapter 8

Question 1

Using the information in Table 3, produce an analysis and a report which establishes the liquidity of the company.

Table 3

Balance sheet	£
Fixed assets	15,400,256
Current assets	
Stocks	4,522,720
Work-in-progress	2,768,934
Debtors	10,720,411
Cash	2,050,189
	20,062,254
Current liabilities	
Creditors	16,146,722
Current taxation	1,283,498
Amount due to Bankers	500,224
Provision for dividends	413,016
	18,343,460
Net current assets	1,718,794
Net total assets	17,119,050

Profit and Loss Account	£
Turnover	70,292,000
Profit before taxation	3,213,424
Taxation	803,354
Profit after taxation	2,410,070
Extraordinary items	210,205
Attributable profit	2,199,865
Dividend	897,492
Transfer to reserves	1,302,373

Question 2

(a) Examine the reasons for using 'ratio analysis' when comparing the financial performance of building companies.
(b) Using the information in Table II, calculate Return on Capital Employed (ROCE), Gearing and Liquidity ratios for the two companies and comment on the results.

Table II (extracts from the Accounts of two building companies)

£ million	COMPANY A			COMPANY B		
	1984	1985	1986	1984	1985	1986
Turnover	94.2	100.0	96.9	1518.0	1581.0	1461.0
Profit	2.7	4.4	2.6	43.6	46.4	48.3
Fixed Assets	60.0	61.2	51.4	340.7	301.8	358.8
Stocks	13.6	13.8	13.4	491.2	487.0	518.0
Work in Progress	9.6	10.3	9.8	61.9	56.1	55.6
Debtors	7.3	8.5	7.3	177.5	199.2	159.4
Cash at bank and in hand	6.7	0.1	0.1	25.7	23.4	18.2
Creditors: due less than one year	32.0	29.6	29.1	453.8	423.8	444.8
: due more than one year	18.8	8.4	6.9	47.2	33.6	44.9
Called up Share Capital	4.1	4.1	4.1	70.4	70.4	70.4
Share Premium	4.2	4.2	4.2	0.0	0.0	0.2
Revaluation Reserve	15.2	21.3	10.1	30.7	9.8	21.0
Capital Reserve	0.1	0.1	0.1	0.0	0.0	0.0
Profit & Loss a/c	22.8	26.2	27.5	379.6	385.9	418.1
Loans	0.0	0.0	0.0	115.3	144.0	110.6

Question 3

(a) Describe the nature and purpose of a share repurchase scheme.
(b) A company wishes to raise additional monies for a capital investment programme in 1992. The Board is considering one of the following alternative methods:
Method X – issuing £125,000 of debentures, paying interest at 10%, with attached warrants enabling the holder of each £1000 debenture to purchase 100 ordinary shares at £2.80 each.

Method Y – making a 'one for three' rights issue at a price of £2.00 per share.

(i) Using the information contained in Table 1, prepare a profit and loss account for 1992 and a balance sheet as at 30 June 1992 for each alternative.
(ii) Comment on the issues to be considered by the Board when deciding which alternative to adopt.

Table 1

Company balance sheet at 30 June 1991

	£		£
Ordinary share capital each @ £1 par value	150,000	Fixed assets	200,000
Revenue reserves	100,000	Stock	70,000
Creditors	80,000	Debtors	60,000
	330,000		330,000

Effect of capital investment scheme

1. Value of fixed assets after investment = £290,000
2. Increase in value of:
   (i) Earnings before interest and tax
   (ii) Creditors
   (iii) Stock
   (iv) Debtors
   } 60%
3. Level of corporation tax 35%
4. Earnings before interest and tax in 1990/91 = £60,000
5. Dividend per share for 1990/91 = 12p

Additional information:

1. Dividend level to be maintained in 1991/92.
2. Debentures and rights issue is made on 1 July 1991.
3. All warrants and rights are exercised during the financial year 1991/92.
4. Excess funds raised are kept in the form of cash.
5. Depreciation charged in year to 30 June 1992 = £10,000.

Question 4

(a) Explain what is meant by 'gearing ratio' in relation to a company's capital structure.
(b) The shareholders of a company have been asked to vote on a proposal to increase long term borrowing limits. Assuming that the cost of capital, after tax, is 8%, calculate the minimum rate of return on capital employed to ensure a 10% net return on equity for each of the following gearing ratios:
   (i) 50%,
   (ii) 70%.

Chapter 12

Question 1

A company is proposing to set up a construction product manufacturing assembly line.

Using the data below, prepare a report, with recommendations, to enable the Board of Directors to make a decision. Give details of the break even point, margin of safety, maximum contribution and profit-volume ratio.

Data	
Maximum capacity	2 000 units
Fixed costs 0–2 000 Units	£100,000.00
Variable costs per Unit	£100.00
Sale price per unit	£200.00

Chapter 13

Question 1

A woodworking company produces a range of products. Using the data of Table 1, and ignoring limiting factors, determine the most beneficial product mix, the associated break even point and the number of units of each product to be produced.

Table 1

Product	Variable Cost	Selling Price	Product Mix				
			1	2	3	4	5
	£	£	%	%	%	%	%
A	35	55	25	–	–	50	–
B	25	49	15	–	60	–	40
C	30	56	10	60	–	30	–
D	65	88	30	40	20	–	30
E	50	73	20	–	20	20	30

FIXED COSTS IN EACH CASE = £496,000

Question 2

(a) Discuss the extent to which 'contribution analysis' can be regarded as an effective method for determining the viability of building projects.

(b) A forecast of the company's activity is shown in Table 2. Comment on this method of analysis and suggest alternatives.

Table 2

	Project				Company Total £m
	A £m	B £m	C £m	D £m	
Project Value	1.20	2.70	1.90	4.20	10.00
Site Expenditure	(1.00)	(2.20)	(1.80)	(3.90)	(8.90)
Apportioned Overheads	(0.10)	(0.22)	(0.18)	(0.39)	(0.89)
Project Profit	0.10	0.28	(0.08)	(0.09)	0.21

Chapter 14

Question 1

Using THREE methods of investment appraisal which do not use discounted cash flow,

(a) Identify the strengths and weaknesses of each appraisal method.
(b) Determine which of the projects listed in Table 1 provide the best investment opportunity, giving reasons for the choice.

Table 1

Project	Initial Capital Cost £000's	Net Annual Revenue £000's			Terminal Value £000's
		Year 1	Year 2	Year 3	
A	200	100	100	100	NIL
B	200	200	20	NIL	NIL
C	200	NIL	100	240	25

Question 2

(a) Discuss the view that discounting methods are better than the payback method for appraising investment in plant and machinery.
(b) The results of an appraisal of three machines are shown in Table 1. Advise on which machine(s) should be purchased, giving reasons for the recommendations and specify any reservations.

Table 1

	Machine			
	X	Y	Z	
Capital Cost £'000	8	10	15	
Resulting net cash inflow				
£'000 in Year 1	4	4	5	
2	4	4	5	
3	1	4	5	
4	–	2	5	
5	–	–	3	
Net Present Value @ 0%	1	4	8	
Net Present Value @ 15%	−022	+1.02	+1.91	
(mid-year discounting)				
Internal Rate of Return	11%	22%	22%	
Payback period in years	2	2½	3	

Question 3

The 'internal rate of return' method of investment appraisal does not rank alternative investment opportunities correctly because it does not take into account the size of the initial investments.

Using the information in Table 1 and Table 2, establish the validity of this statement.

Table 1

Project cash flow

Year	Type of cash flow	Project A cash flow £	Project B cash flow £
0	capital investment	1 000	10 000
1	income	250	2 300
2	income	300	3 200
3	income	390	3 800
4	income	460	4 200

NB: Cost of capital = 10%

Table 2

The present value of £1

Year	10%	15%
1	0.909	0.870
2	0.826	0.756
3	0.751	0.658
4	0.683	0.572

Chapter 15

Question 1

Using the information in Table 2, calculate the anticipated profit margin for 75%, 90% and 100% levels of activity and comment on your results.

Table 2

Item	Level of Activity		
	100%	90%	75%
Number of units sold per annum	100	90	75
Sale price per unit (£000's)	60	57	54
Prime cost per unit (£000's)	40	40	40
Variable overhead (% of sale price)	2.5	2.5	2.5
Semi-variable overhead (% of prime cost)	10	11	12
Fixed overhead per annum (£000's)	800	800	800

NB: Output at 100% level of activity = 100 units

Question 2

The following information relates to one month's activity on a building project:

	Budget (£000's)	Actual (£000's)
Value of work completed	50.4	46.7
Materials used	24.8	23.9
Direct labour	5.2	4.8
Sub-contract labour	16.9	16.7

Material prices are 7% above the budget level. Direct labour is 5% above the budget level. Sub-contract labour is 3% below the budget level.

Determine the true variances for the month.

Chapter 16

Question 1

(a) Explain the differences between 'cash flow' and 'working capital' and comment on their significance in the operation of building companies.
(b) Using the information in Table 1, Calculate the bank balance of a building company at the end of period 7 and comment on its financial situation.

Table 1

(An extract of key financial forecasts)

Period	1	2	3	4	5	6	7
	£ million						
Sales	2.2	2.5	2.7	2.5	2.9	2.6	2.4
Total operating costs	2.1	2.2	2.2	2.3	2.4	2.3	2.1
Profit	.1	.3	.4	.3	.5	.3	.3
Revenue receipt	2.0	2.4	2.3	2.4	2.4	2.6	2.6
Capital receipts	.1	.2	.0	.0	.1	.0	
Revenue payments	1.9	2.0	1.8	2.1	2.2	2.3	2.4
Capital payments	.3	.2	.4	.4	.2		
Capital purchases	.4	.2	.1	.0	.0	.0	.0
Capital sales	.1	.0	.2	.0	.0	.1	.0

Opening bank balance 0.2

Question 2

Prepare a cash budget for the quarter 1 July to 30 September 1992, using the receipts and payments method and the information in Table 4.

Determine the action which the company should take.

Table 4 – Requirements for cash budget

Month	Sales	Materials	Labour	Overheads			
				Building Production	Sales	Design	Administration
	£	£	£	£	£	£	£
April	240 000	70 000	120 000	24 000	4 800	2 000	25 000
May	300 000	100 000	140 000	36 000	6 000	2 000	25 000
June	480 000	160 000	230 000	58 000	9 600	2 000	30 000
July	600 000	200 000	280 000	71 000	12 000	2 000	35 000
August	500 000	160 000	240 000	58 000	10 000	2 000	30 000
September	240 000	70 000	120 000	24 000	4 800	2 000	25 000
October	120 000	40 000	60 000	16 000	2 000	2 000	25 000

Additional information

(i) Cash balance at 1 July is expected to be = £80 000

(ii) Period of credit allowed by creditors = 1 month

(iii) Period of credit allowed to debtors = 3 months

(iv) Delay in payment of overheads = 1 month

(v) Delay in payment of labour = 14 days

(vi) £0.30 call on ordinary share capital of one million £1 shares is due on 1 September 1992.

Question 3

A building company will be undertaking a project with the estimated outputs and inputs shown in Table 1.

Table 1

Month	Ouput (£000's)	Input (£000's)
1	100	90
2	200	180
3	300	270
4	300	270
5	300	270
6	300	270
7	200	180
8	100	90
	1,800	1,620

(i) The client will make payments monthly in arrears taking a 5% retention from the value of the output.
(ii) The retention monies will be released 3 months after the end of the contract following a satisfactory survey.
(iii) Inputs will be paid for on a continuous basis during each month.
(iv) Finance for the project will be obtained by using an overdraft facility charging 15% per annum.

Calculate the maximum overdraft required and the interest payable and comment on the ways in which the interest payment could be reduced.

Question 4

Table 3 is a Cash Flow Assessment Chart which shows predicted values for a project as follows:

Project	
Value	£48 000
Duration	6 months
Retention	5%
Mark up	10%
Defects liability period	6 months
Certificates	monthly
Payments	one month after the issue of certificates
Contractor's cost of capital	18% per annum

Work has proceeded in accordance with the predicted plan but during a budget monitoring exercise the following discrepancies were identified.

(i) Certificate numbers 1 & 2 showed a 10% undervaluation and number 4 a 75% undervaluation.
(ii) Contract retention was only 3%.

Using the Worksheet MII/91/5/A, calculate the effect of these discrepancies on contract cost and comment upon it.

Table 3

CASH FLOW ASSESSMENT CHART

	A	B	B – profit (B × 100/110)	B – 5% ret	–ve shortfall	A–5% ret	GROSS
Month	Monthly Budget retention) £	Cumulative Budget £	Contract Expenditure £	Money received (cumulative- £	Net cash flow £	Net monthly £	Max cash £
1	3 750	3 750	3 409	–	– 3 409	–	– 3 409
2	8 250	12 000	10 909	3 562	– 7 347	3 562	– 10 909
3	12 000	24 000	21 818	11 400	– 10 418	7 838	– 18 256
4	12 000	36 000	32 727	22 800	– 9 927	11 400	– 21 327
5	8 250	44 250	40 227	34 200	– 6 027	11 400	– 17 427
6	3 750	48 000	43 636	42 038	– 1598	7 838	– 9 436
				45 600 + 1 200 ret	+ 1964 + 1 200	3 562 + 1 200 ret	– 1 598
				46 800 + 1 200 ret	+ 1 200	+ 1 200	
				48 000		48 000	

Worksheet MII/90/5

The Chartered Institute of Building

BUILDING ECONOMICS AND FINANCE – WORKSHEET CASH FLOW ASSESSMENT CHART

Month	A	B	Actual Contract Expenditure	Actual money received (B – retention	–ve shortfall	Actual monthly income (A – retention)	Gross
	Monthly Budget* £	Cumulative Budget* £	£	£	Net cash flow £	£	Max cash requirement £
1			3 409	–	–3 409	–	–3 409
2			10 909				
3	13 200	24 000	21 818				
4			32 727				
5	17 250	44 250	40 227				
6	3 750	48 000	43 636				
7							

*Adjusted for undervaluation

Question 5

Sheets MII/90/5/A and B and the graph, when completed, must be handed in with the answer script).

Using the information given in Table 2 and on Sheet MII/90/.5/A, complete the average payment delay statement, prepare a cost time graph and finalise the project cash flow statement on Sheet MII/90/5/B. Comment on the results obtained.

Table 2

Cost Centre	Proportion of total cost		Payment Delay
	Months 1–3 (%)	Months 4–6 (%)	(No of weeks)
Labour	10	25	1
Material	15	40	4
Plant	60	15	5
Sub-Contractors	15	20	2

(i) Payment of certificates takes place on the first day of the next month after the end of the month in which payment is raised,
(ii) Mark up = 10%,
(iii) Retention = 5%,
(iv) Defects liability period = 6 months.

MII/90/5/A

BUILDING ECONOMICS AND FINANCE – INFORMATION AND WORKSHEET

Project value, cost and payment analysis

Month End	Cumulative value £	Cumulative cost £	Cumulative Payment £
May	81 000	73 636	–
June	232 000	210 909	–
July	358 000	325 455	76 950
August	554 000	503 636	220 400
September	634 000	576 364	340 100
October	700 000	636 364	526 300
November	–	–	602 300
December	–	–	682 500
June	–	–	700 000

Average payment delay statement

Input	Months 1–3		Months 4–6
Labour	10% × 1 week	= 0.10	
Materials	15% × 4 weeks	= 0.60	
Plant	60% × 5 weeks	= 3.00	
Sub-contractors	15% × 2 weeks	= 0.30	
	weighted delay	=	weighted delay =

MII/90/5/B

BUILDING ECONOMICS AND FINANCE – CASH FLOW WORKSHEET

Cash flow statement

Month	Cost		Income		Cash flow position	
	Beginning	End	Beginning	End	Beginning	End
May	–		–	–	–	
June		73 636	–	–		−73 636
July	73 636	210 909	76 950	76 950		
August	210 909		220 400	220 400		
September	325 455			340 100		
October						
November		636 364				
December	636 364	636 364	682 500			
June	636 364	636 364	700 000	–		+63 636

Chapter 17

Question 1

Using information given in Table 2, calculate the variances for labour cost, labour rate and labour efficiency.

Table 2

Information supplied by estimating department

Labour constant:	$4m^2$ per man hour
'All-in' labour rate:	£4.80 per hour

Information gained from weekly time and remeasurement sheets

Total production	$360m^2$
Hours worked by operatives:	39 hours each
Gang Size:	2 operatives
'All-in' labour rate:	£5.00 per hour

SOLUTIONS TO QUESTIONS

Chapter 8

Question 1

Solvency and liquidity are topics covered in Section 8.7. The company's current ratio is calculated as:

$$\frac{\text{Current assets}}{\text{Current liabilities}} : 1 = \frac{20\,062\,254}{18\,343\,460} : 1 = 1.09 : 1$$

This discloses that short-term liabilities are covered, but only just, by current assets. This, however, is too crude a measure to be of great value. A more refined version has to be calculated because the stocks and work-in- progress are not a form of asset which can be converted into cash at short notice. A calculation should be used which includes only items actually in cash or in a form capable of being converted into cash quickly.

Assuming that all sales are on credit, the average length of time taken by debtors can be calculated as:

$$\frac{\text{Debtors} \times 365}{\text{Turnover}} = \frac{10\,720\,411 \times 365}{70\,292\,000} = 55.7\,days$$

On average, therefore, debtors take about 8 weeks to settle their accounts, and this is regarded as reasonable. Debtors can therefore be regarded as being convertible quickly and as such are a component of the previously mentioned refined calculation, known as the quick, acid test or liquidity ratio:

$$\frac{\text{Quick assets}}{\text{Current liabilities}} : 1 = \frac{(10\,720\,411 + 2\,050\,189)}{18\,343\,460} : 1 = 0.70 : 1$$

It is apparent that the company is unable to meet its external obligations due for settlement in the near future (out of its cash and near cash resources), by a substantial margin. This is a very serious situation: the company could be declared insolvent at any time, be placed in receivership and/or wound up.

Question 2

(a) Ratio analysis is used for the comparison of the financial performance of building companies because, as for other companies, ratios enable the results of different sizes of businesses to be reduced to a common base. This is the effect of eliminating differences attributable to uneven scales of operating activity.

(b) For each of the calculations, the figures given in the question for the two companies need to be rewritten in such a way that sub-totals can be derived. These can then be inserted in the various formulae from which the ratios are derived. The formulae are to be found in Chapter 8.

	Company A			Company B		
	Year 4 £m	Year 5 £m	Year 6 £m	Year 4 £m	Year 5 £m	Year 6 £m
Fixed assets	60.0	61.2	51.4	340.7	301.8	358.8
Current assets						
Stocks	13.6	13.8	13.4	491.2	487.0	518.0
Work in progress	9.6	10.3	9.8	61.9	56.1	55.6
Debtors	7.3	8.5	7.3	177.5	199.2	159.4
Cash and bank	6.7	0.1	0.1	25.7	23.4	18.2
	37.2	32.7	30.6	756.3	765.7	751.2
Creditors due in −1 year	32.0	29.6	29.1	453.8	423.8	444.8
Net current assets	5.2	3.1	1.5	302.5	341.9	306.4
Total assets less current liabilities	65.2	64.3	52.9	643.2	643.7	665.2
Creditors due in +1 year (including loans)	18.8	8.4	6.9	162.5	177.6	155.5
Called-up share capital	4.1	4.1	4.1	70.4	70.4	70.4
Reserves						
Share premium	4.2	4.2	4.2	–	–	0.2
Revaluation reserve	15.2	21.3	10.1	30.7	9.8	21.0
Capital reserve	0.1	0.1	0.1	–	–	–
Profit and loss	22.8	26.2	27.5	379.6	385.9	418.1
Shareholders' funds	46.4	55.9	46.0	480.7	466.1	509.7
Capital employed	65.2	64.3	52.9	643.2	643.7	665.2
Turnover	94.2	100.0	96.9	1518.0	1581.0	1461.0
Net profit after tax	2.7	4.4	2.6	43.6	46.4	48.3

	Company A Year 4	Company A Year 5	Company A Year 6	Company B Year 4	Company B Year 5	Company B Year 6
Return on capital employed $\frac{\text{Net profit}}{\text{Capital employed}} \times \frac{100}{1}$	$\frac{2.7}{65.2} \times \frac{100}{1}$ = 4.1%	$\frac{4.4}{64.3} \times \frac{100}{1}$ = 6.8%	$\frac{2.6}{52.9} \times \frac{100}{1}$ = 4.9%	$\frac{43.6}{643.2} \times \frac{100}{1}$ = 6.8%	$\frac{46.4}{643.7} \times \frac{100}{1}$ = 7.2%	$\frac{48.3}{665.1} \times \frac{100}{1}$ = 7.3%
This is the product of $\frac{\text{Net profit}}{\text{Turnover}} \times \frac{100}{1}$	$\frac{2.7}{94.2} \times \frac{100}{1}$ = 2.9%	$\frac{4.4}{100} \times \frac{100}{1}$ = 4.4%	$\frac{2.6}{96.9} \times \frac{100}{1}$ = 2.7%	$\frac{43.6}{1\,518.0} \times \frac{100}{1}$ = 2.9%	$\frac{46.4}{1\,581.0} \times \frac{100}{1}$ = 2.9%	$\frac{48.3}{1\,461.0} \times \frac{100}{1}$ = 3.3%
$\frac{\text{Turnover}}{\text{Capital employed}}$	$\frac{94.2}{65.2}$ = 1.4	$\frac{100}{64.3}$ = 1.6	$\frac{96.9}{52.9}$ = 1.8	$\frac{1\,518.0}{643.2}$ = 2.4	$\frac{1\,581.0}{643.7}$ = 2.5	$\frac{1\,461.0}{665.1}$ = 2.2
Gearing $\frac{(\text{Long-term loans} \times 100)}{\text{Shareholders' funds}}$	Nil	Nil	Nil	$\frac{(115.3 \times 100)}{(480.7 + 115.3)}$ = 19.3%	$\frac{(144.0 \times 100)}{(466.1 + 144.0)}$ = 23.6%	$\frac{(110.6 \times 100)}{(509.7 + 110.6)}$ = 17.8%
Liquidity $\frac{\text{Current assets}}{\text{Current liabilities}} : 1$	$\frac{37.2}{32.0}$ = 1.2:1	$\frac{32.7}{29.6}$ = 1.1:1	$\frac{30.6}{29.1}$ = 1.1:1	$\frac{756.3}{453.8}$ = 1.6:1	$\frac{765.7}{423.8}$ = 1.8:1	$\frac{751.2}{444.8}$ = 1.7:1
$\frac{\text{Quick assets}}{\text{Current liabilities}} : 1$	$\frac{14.0}{32.0}$ = 0.4:1	$\frac{8.6}{29.6}$ = 0.3:1	$\frac{7.4}{29.1}$ = 0.2:1	$\frac{203.2}{453.8}$ = 0.4:1	$\frac{222.6}{423.8}$ = 0.5:1	$\frac{177.6}{444.8}$ = 0.4:1
$\frac{(\text{Debtors} \times 365)}{\text{Turnover}}$	$\frac{(7.3 \times 365)}{94.2}$ = 28 days	$\frac{(8.5 \times 365)}{100.0}$ = 31 days	$\frac{(7.3 \times 365)}{96.9}$ = 27 days	$\frac{(177.5 \times 365)}{1\,518.5}$ = 42 days	$\frac{(199.2 \times 365)}{1\,581.0}$ = 46 days	$\frac{(159.4 \times 365)}{1\,461.0}$ = 40 days

Return on capital employed (ROCE)

Company B has the higher ROCE throughout the three years but both companies show a steady improvement. The reason behind Company B's superiority is its ability to utilise its capital employed much more effectively than can Company A.

Gearing

Company A has no long-term borrowings; at around 20 per cent B's gearing is well within acceptable limits.

Liquidity

Company B's current ratio is stronger than Company A's but both companies have quick ratios at a crisis level, indicating that they are desperately short of liquid funds. There is a marked difference in the speed with which they collect their debts. Company B could bring about some improvement in its liquid position by granting less generous credit terms (although this might result in consumer resistance) and/or exercising stricter credit control.

The problem is more fundamental than could be rectified by adjustments of this sort, however. A substantial injection of liquid funds is needed by both companies. This could take the form of additional share issues and/or long-term borrowings, both of which would appear to be attractive to investors on the basis of the companies' performance.

Lack of sufficient detailed information prevents further comments being made.

Question 3

(a) Under the Companies Act 1985 companies are permitted to repurchase (buy back) their own shares on the open market and cancel them, subject to the observance of certain conditions. The purpose of this manoeuvre is to remove excess funding from the company when it has sufficient funding from internally generated ploughed back profits. In a private limited company this also enables a shareholder to dispose of his or her shares, because, as explained in Chapter 1, Section 1.3, shares in a private company cannot be sold to the public.

(b) *Profit and loss account (extract) for the year ended 30 June 1992*

	Method	
	X	*Y*
Profit before interest and tax	96 000	96 000
Debenture interest (10% × 125 000)	(12 500)	–
Profit before tax	83 500	96 000
Corporation tax (35%)	(29 225)	(33 600)
Profit after tax	54 275	62 400
Dividends (12 pence per share)	(19 500)	(24 000)
Retained profit: for year	34 775	38 400
Brought forward	100 000	100 000
Carried forward	134 775	138 400

Balance sheet as at 30 June 1992

Workings		*Method X* £	*Method Y* £
	Fixed assets	290 000	290 000
	Current assets		
	Stock	112 000	112 000
	Debtors	96 000	96 000
W2	Bank	74 775	18 400
		282 775	226 400
	Creditors due in less than one year	128 000	128 000
	Net current assets	154 775	98 400
	Total assets less current liabilities	444 775	388 400
	Creditors due in more than one year		
	10% debentures	125 000	–
	Net assets	319 775	388 400
	Share capital and reserves		
W1	*Called-up share capital*	162 500	200 000
	Reserves		
W2	Share premium	22 500	50 000
Profit and loss	Revenue reserves	134 775	138 400
	Shareholders' funds	319 775	388 400

Workings		*Method X* £	*Method Y* £
W1	*Ordinary share capital*		
	Balance 30 June 1991	150 000	150 000
	Conversion of warrants [(125 000/1 000) × 100 × £1.00 nominal]	12 500	
	Rights issue [(150 000/3) × £1.00 nominal]		50 000
	Balance 30 June 1992	162 500	200 000
	Share premium (see Chapter 4, Section 4.3)		
	Conversion of warrants [(125 000/100)x100x(£2.80 – £1.00)]	22 500	
	Rights issue [(150 000/3) × (£2.00 – £1.00)]		50 000

W2 The bank balances can be derived as the only missing figures in the balance sheet but can also be built up:

	Method X £	Method Y £
Receipts		
10% debenture issue	125 000	
Conversion of warrants (12 500 + 22 500)	35 000	
Rights issue (50 000 + 50 000)		100 000
Debtors	60 000	60 000
Net receipts from trading (profit and loss)	34 775	38 400
	254 775	198 400
Payments		
Creditors	80 000	80 000
Fixed assets [290 000 + 10 000 (depreciation) – 200 000)	100 000	100 000
	180 000	180 000
Excess of receipts – bank balance	74 775	18 400

In view of the lack of precise information, figures for receipts from debtors and from trading and payments to creditors are netted-off figures, but the closing balances are correct. No opening bank balance was given; it has to be assumed that there must have been an overdraft included in 'creditors due within one year'.

Issues considered by the Board would be whether future profits would be adequate to meet interest payments, whether dividend rates could be maintained on the enlarged share capital; the effect on gearing; and the reasons for raising £125 000 of debentures bearing 10 per cent interest when this will result in a cash surplus of nearly £75 000 which will not earn any interest.

Question 4

(a) There are various definitions of gearing ratio, but one that is commonly accepted is that it shows the relationship between borrowed funds and total funds, as described in Chapter 8, Section 8.8.

(b)

	(i) %		%	*(ii)* %		%
Total funds	100			100		
less gearing	50 × 8%	=	4.0	70 × 8%	=	5.6
Equity funds	50 × 10%	=	5.0	30 × 10%	=	3.0
Minimum rate of return on capital employed			9.0			8.6

This can be proved, using a capital employed figure of £4 000 000.

	Gearing (%)	*Capital employed £*	*Return (%)*	*Return (£)*	*Gearing (%)*	*Capital employed £*	*Return (%)*	*Return (£)*
Borrowing	50	2 000 000	8.0	160 000	70	2 800 000	8.0	224 000
Equity	50	2 000 000	10.0	200 000	30	1 200 000	10.0	120 000
Total funds	100	4 000 000	9.0	360 000	100	4 000 000	8.6	344 000

Chapter 12

Question 1

Contribution (C) per unit = Sales price minus variable cost
= £200 − £100
= £100

Profit/volume ratio (P/V)

$$= \frac{\text{(Sales price minus variable cost)}}{\text{Sales price}}$$

$$= \frac{(200-100)}{200} = \frac{100}{200}$$

= 0.5

Break even point (B/E)
In quantity

$$= \frac{\text{Fixed costs}}{\text{Contribution}}$$

$$= \frac{£100\,000}{£100}$$

= 1000 units

In value

$$= \frac{\text{Fixed costs}}{\text{P/V ratio}}$$

$$= \frac{£100\,000}{0.5}$$

= £200 000

Margin of safety = Maximum capacity minus B/E quantity
= 2000 − 1000
= 1000 units

Margin of safety %

$$= \frac{\text{Margin of safety}}{\text{Maximum capacity}} \times \frac{100}{1}$$

$$= \frac{1\,000}{2\,000} \times \frac{100}{1}$$

= 50%

Maximum contribution = Maximum capacity × contribution
= 2000 × £100
= £200 000

The report should state that all the indicators are favourable:

- a large contribution able to absorb the fixed costs and leave a substantial profit;
- low commercial risk from the relatively low break-even point and the wide margin of safety; and
- profitable product as shown by reasonably high P/V ratio.

Chapter 13

Question 1

	Product					
	A	*B*	*C*	*D*	*E*	*Total*
	£	£	£	£	£	£
Selling price	55	49	56	88	73	–
Variable cost	35	25	30	65	50	–
Contribution	20	24	26	23	23	–

Product mix	%	£	%	£	%	£	%	£	%	£	*Total contribution* £
1	25	5.0	15	3.6	10	2.6	30	6.9	20	4.6	22.7
2	–		–		60	15.6	40	9.2	–		24.8
3	–		60	14.4	–		20	4.6	20	4.6	23.6
4	50	10.0	–		30	7.8	–		20	4.6	22.4
5	–		40	9.6	–		30	6.9	30	6.9	23.4

In the lower half of the table the given percentages for the mix are applied to the contributions per product derived in the upper half. From the calculations it can be see that Mix 2 produced the highest weighted contribution (£24.80).

Break-even point (quantity):

$$\frac{\text{Fixed costs}}{\text{Contribution}} = \frac{£496\,000}{£24.80} = 20\,000 \text{ units}$$

Production quantities at break-even point (Product Mix 2)

Product	Mix	Quantity units	Sales		Variable cost		Contribution
			Price	Amount	Per unit	Amount	
	%	No	£	£	£	£	£
C	60	12 000	56	672 000	30	360 000	312 000
D	40	8 000	88	704 000	65	520 000	184 000
	100	20 000		1 376 000		880 000	496 000
						Fixed costs	(496 000)
						Break even	–

Question 2

(a) The rationale behind contribution analysis is that, by comparing the costs directly attributable to a project with its (market) value it is readily apparent whether the project is viable or not. A positive contribution – the extent of project value in excess of direct costs – indicates that a project is a viable proposition; a negative contribution, resulting from direct costs exceeding the project's value, shows that the project is not viable. However, in the latter case, the company might still be prepared to undertake the project, knowing that a loss will result, in order to achieve some other objective. For example, the objective might be to secured a foothold in a particular locality or sector.

In contribution analysis, fixed costs are dealt with in total only and are not attributed to projects. This treatment is appropriate because, by definition, fixed costs will not be affected by the decision to undertake a project or not, and as such are irrelevant.

(b) In Table 2 of the question all costs have been attributed to projects, as a result of which, Projects C and D are shown to produce losses. On this basis, presumably, they would not be undertaken. The answer to Question 2(a) above indicates that the Table 2 approach is fallacious. Any decision should result from a contribution analysis approach which requires that Table 2 be rewritten:

	Project				
	A £m	B £m	C £m	D £m	Total £m
Project value	1.20	2.70	1.90	4.20	10.00
Site expenditure (direct costs)	(1.00)	(2.20)	(1.80)	(3.90)	(8.90)
Contribution	0.20	0.50	0.10	0.30	1.10
less Apportioned overheads					(0.89)
Profit					0.21

It is now apparent that all four projects have positive contributions and are therefore financially viable. If Projects C and D were to be abandoned, the whole of the fixed costs would have to be borne by A and B:

	£m
Contribution	
A	0.20
B	0.50
	0.70
Less fixed overheads	(0.89)
Loss	(0.19)

There would therefore be an overall loss of £0.19m, which can be directly related to the absence of a contribution from C and D:

	£
Original profit	0.21
Loss from undertaking A and B (above)	(0.19)
Deterioration in financial position	(0.40)
equal to	
Loss of contribution	
C	0.10
D	0.30
	0.40

Chapter 14

Question 1

(a) Three methods of investment which do not use discounted cash flow are:

- payback;
- accounting rate of return:
  based on average investment; and
  based on initial investment.

Strengths and weaknesses of each of these methods have been identified in Chapter 14, Sections 14.4 and 14.5, to which reference should be made at this point to avoid repetition.

(b) Payback

		Project	
	A	*B*	*C*
	£000	*£000*	*£000*
Initial cost	200	200	200
Net annual revenue to payback			
Year 1	100	200	–
Year 2	100	–	100
Year 3	–	–	100
	200	200	200
Payback period	2 yrs	1 yr	2 yrs 5 mths

On the basis of payback, Project B, having the shortest payback period, would be selected.

Accounting rate of return

The initial workings are identical for each of the two available methods under this heading. First, a profit and then an average profit has to be calculated by converting the net annual revenue figures into profit figures. This is achieved by charging depreciation against the net annual revenue figures which, in effect, are net cash inflows. Second, the figures calculated are then used to calculate the respective rates of return.

		Project	
	A	B	C
	£000	*£000*	*£000*
Net annual revenue (net cash inflow) total	300	220	340
less Depreciation (initial cost less terminal value)	200	200	175
Total profit	100	20	165
Average annual profit (3 years life)	33	7	55

Accounting rate of return on average investment

Average investment = (initial cost plus terminal value)/2

		Project	
	A	B	C
Average annual profit (£000)	33	7	55
Average investment (£000)	100	100	112
	[(200 + 0)/2]	[(200 + 0)/2]	[(200 + 25)/2]
Rate of return (%)	33	7	49

In contrast to the results obtained from payback, Project B, which had the shortest payback period, now has the lowest rate of return; and Project C has the highest rate of return but the longest payback period.

Accounting rate of return on initial investment

		Project	
	A	B	C
Average annual profit (£000)	33	7	55
Initial investment (£000)	200	200	200
Rate of return (%)	16.5	3.5	27.5

Project C emerges again as the clear leader.

General

Project C presents the best investment opportunity and would ordinarily be selected. However, if the overriding consideration is the speediest recoupment of the initial outlay, the choice would fall on Project B.

Chapter 14

Question 2

(a) The shortcomings of the payback method were dealt with in Chapter 14, Section 14.4, together with its advantages. You should refer to that for this part of the answer. Whereas payback takes no account of cash flow once payback has been reached, discounting methods use cash flows over the whole life of the project and at the same time recognise that money has a time value, as described in Chapter 14, Section 14.6.

(b) If the payback method was being employed as the basis of selection, Machine X would be chosen, but this machine would be firmly rejected if a discounting method were the discriminator. It has a negative net present value, indicating that it fails to meet the minimum 15 per cent discounted return; indeed, as shown by the internal rate of return figure, its yield is 11 per cent.

The choice therefore lies between Machines Y and Z. It would seem that Z, having the higher net present value (+1.91) is better than Y (+1.02). This fails to take into account the fact that initial investment is different for each machine. In this situation, Chapter 14, Section 14.7 stated that an NPV index should be calculated:

$$\frac{\text{Present value}}{\text{Initial outlay}}$$

giving the following results:

$$\text{Machine Y} \quad \frac{(10 + 1.02)}{10} = 1.102$$

$$\text{Machine Z} \quad \frac{(15 + 1.91)}{15} = 1.127$$

The results are now so close as to be almost indistinguishable. This is confirmed by the identical yields (22 per cent). However, Machine Y might be chosen because of its faster payback combined with its lower initial outlay (capital cost).

Question 3

In order to test this statement it is necessary to appraise the projects on the internal rate of return (yield) method and on those other methods which take size of initial investment into account; namely, net present value and accounting rate of return. There is, however, not enough information available for this last named method to be applied to the given data. The net present value and the internal rate of return methods are applied as explained and illustrated in Chapter 14, Sections 14.7 and 14.8. Net present values are calculated in the schedule which now follows:

Year	P/V factor (10%)	Project A Cash flow £	Project A Present value £	Project B Cash flow £	Project B Present value £
0	1.000	(1 000)	(1 000)	(10 000)	(10 000)
1	0.909	250	227	2 300	2 091
2	0.826	300	248	3 200	2 643
3	0.751	390	293	3 800	2 854
4	0.683	460	314	4 200	2 869
Net present value			82		457
	(15%)				
0	1.000	(1 000)	(1 000)	(10 000)	(10 000)
1	0.870	250	218	2 300	2 001
2	0.756	300	227	3 200	2 419
3	0.658	390	257	3 800	2 500
4	0.572	460	263	4 200	2 402
Net present value			(35)		(678)

In order to calculate the yield (internal rate of return) of each project, the above data could be plotted on a graph, as illustrated in Chapter 14, Section 14.8, Figure 14.2 but it is simpler in this case to use the formula given in that same section.

$$A + \left(\frac{(B - A)}{1} \times \frac{a}{(a + b)}\right)$$

The meaning of the symbols is as given in Chapter 14, Section 14.8.

Project A

$$10 + \left(\frac{(15 - 10)}{1} \times \frac{82}{(82 + 35)}\right) = 10 + \left(\frac{5}{1} \times \frac{82}{117}\right)$$

$$= 10 + 3.5 = 13.5\%$$

Project B

$$10 + \left(\frac{(15 - 10)}{1} \times \frac{457}{(457 + 678)}\right) = 10 + \left(\frac{5}{1} \times \frac{457}{1\,135}\right)$$

$$= 10 + 2.0 = 12.0\%$$

As the figures stand at the moment, Project B would be selected on the basis of its higher net present value, and Project A on its higher yield.

Further calculations need to be carried out on the net present value figures because of the disparity in size of the initial capital investment. Chapter 14, Section 14.7 states that this is done by indexing:

Project A

$$\frac{(1\,000 + 82)}{1\,000}$$

$$= 1.08$$

Project B

$$\frac{(1\,000 + 457)}{1\,000}$$

$$= 1.05$$

On this basis, Project A would be selected on the grounds of its higher index.

This result agrees with that of the yield method. The validity of the statement that the internal rate of return method does not rank alternative investment opportunities correctly because it ignores the size of the initial investments is not borne out by the above figures.

Chapter 15

Question 1

The subject of budget preparation for different activity levels was the subject matter of Chapter 15, Section 15.6. The figures given in the question can be rescheduled to provide the requisite information.

		Activity level	
	100%	*90%*	*75%*
Sales (units)	100	90	75
	£000	*£000*	*£000*
Sales	6 000.00	5 130.00	4 050.00
Prime cost	4 000.00	3 600.00	3 000.00
Variable overhead	150.00	128.25	101.25
Semi-variable overhead	400.00	396.00	360.00
	4 550.00	4 124.25	3 461.25
Fixed costs	800.00	800.00	800.00
	5 350.00	4 924.25	4 261.25
Profit/(loss)	650.00	205.75	(211.25)

The company should ensure that the activity level does not fall much below 90 per cent, otherwise a loss will be incurred.

Question 2

The variances between the actual and budgeted figures as given are not a valid representation of the situation because they are at two different activity levels. True variances can be obtained by flexing the budget, as explained in Chapter 15, Sections 15.5 and 15.6

Actual activity is (46.7 × 100)/50.4, that is, 92.65 per cent of budgeted activity. The budget is flexed to this level and the actual figures can then be compared.

	Flexed budget *£m*	*Actual* *£m*	*(True) Variance* *£m*
Materials (92.65% × 24.8)	23.0	23.9	0.9 (A)
Direct labour (92.65% × 5.2)	4.8	4.8	–
Sub-contract (92.65% × 16.9)	15.7	16.7	1.0 (A)

Using the other information given, the variances can be further analysed.

	Variance *£m*
Materials used	
Price variance (7/107 × 23.9)	0.2 (A)
Usage variance (balance)	0.7 (A)
Total variance	0.9 (A)
Direct labour	
Rate variance (5/105 × 4.8)	0.2 (A)
Efficiency variance (balance)	0.2 (F)
Total variance	–
Sub-contract	
Rate variance (3/97 × 16.7)	0.5 (F)
Efficiency variance (balance)	1.5 (A)
Total variance	1.0 (A)

Chapter 16

Question 1

(a) Cash flow is the term given to the effect of capital and revenue receipts and payments in a business. Working capital is the aggregate of current assets minus current liabilities.

(b) Preparation of phased cash budgets is dealt with in Chapter 16, Section 16.2. Figures for sales, operating costs and profit are not relevant to the budget. Only items representing actual cash flows are used.

	Cash budget						
	1	2	3	4	5	6	7
	£m	£m	£m	£m	£m	£m	£m
Receipts							
Capital	0.1		0.2			0.1	
Revenue	2.0	2.4	2.3	2.4	2.4	2.6	2.6
	2.1	2.4	2.5	2.4	2.4	2.7	2.6
Payments							
Capital	0.3	0.2	0.4	0.4	0.2		
Revenue	1.9	2.0	1.8	2.1	2.2	2.3	2.4
	2.2	2.2	2.2	2.5	2.4	2.3	2.4
Surplus/(deficit)	(0.1)	0.2	0.3	(0.1)	–	0.4	0.2
Opening balance	0.2	0.1	0.3	0.6	0.5	0.5	0.9
Closing balance	0.1	0.3	0.6	0.5	0.5	0.9	1.1

There has been a rapid and substantial improvement in the cash position.

Question 2

The cash budget is prepared on a phased basis allowing for time lags as in Chapter 16, Section 16.2.

	Cash budget		
	July	August	September
	£	£	£
Receipts			
Debtors	240 000	300 000	480 000
Share call			300 000
	240 000	300 000	780 000
Payments			
Materials	160 000	200 000	160 000
Labour	115 000	140 000	120 000
	140 000	120 000	60 000
Overheads			
Building production	58 000	71 000	58 000
Sales	9 600	12 000	10 000
Design	2 000	2 000	2 000
Administration	30 000	35 000	30 000
	514 600	580 000	440 000
Surplus/(deficit)	(274 000)	(280 000)	340 000
Opening balance	80 000	(194 600)	(474 600)
Closing balance	(194 600)	(474 600)	(134 600)

It is apparent that the company needs to arrange a massive overdraft facility of just under £500 000. This could be limited to under £200 000 if the share call was advanced to July.

Question 3

Calculation of overdraft requirement

Month	*Receipts less 5% retention time-lagged by 1 month*	*Payments*	*Balance ()overdraft*	*Overdraft interest (1/12 × 15% per month)*
	£000	*£000*	*£000*	*£000*
1	–	90	(90)	1.125
2	95	180	(175)	2.188
3	190	270	(255)	3.187
4	285	270	(240)	3.000
5	285	270	(225)	2.813
6	285	270	(210)	2.625
7	285	180	(105)	1.312
8	190	90	(5)	0.063
9	95	–	90	–
10	–	–	90	–
11 Retentions released	90	–	180	–
	1800	1620		16.313

The maximum is required in Month 3. There is very little scope for reducing the overdraft interest except by negotiating credit terms with the creditors for inputs. Acceleration of receipts does not seem to be a feasible possibility.

Question 4

This is a question involving assessment of cash requirements by means of proformae. A much simpler version of this, given in Chapter 16, Section 16.3, to which you should now refer, could have been used to give the same information as the needlessly complex method used in the question. As originally set, the question contained various errors which I have corrected. In view of the complications this solution starts by explaining how the given figures were calculated.

Contract expenditure column

Figures have to be derived by eliminating the profit element of the cumulative budget (value of completed work). Mark up is given as 10 per cent, or 1/10. For reasons given in Chapter 5, Section 5.6, this has to be converted to a margin figure:

$$1/(10 + 1) = 1/11$$

If the profit element of the budget is 1/11, it follows that the cost element is 10/11. Hence, contract cost expenditure is 10/11 of cumulative budget.

Money received

Ninety-five per cent of the value of the cumulative budget is in fact received to begin with (the remaining 5 per cent being retained until after completion) and is time lagged.

Net cash flow

This is contract expenditure minus money received.

Net monthly income

This is the current month component of money received.

Maximum cash requirement

The effect of the time lag is that money due for Month 2 work is due one Month after the issue of the certificate: that is, at the end of Month 3. In the meantime, Month 3 contract expenditure has to be met. The result is that, although at the end of Month 3 the net cash flow is −£10 418, the maximum cash requirement during that month is −£18 256, until it is reduced by £7 838 to −£10 418 at the end of Month 3.

Solution

Once the principles on which the question has been compiled have been understood, they have to be applied, on the basis of the changed circumstances, to the proforma.

After this has been completed it is then possible to calculate the effect on contract cost, using the interest rate of 18 per cent per annum. The result is additional interest of £118 (£1 354 − £1 236).

Maximum cash requirement and interest

Month	*Planned requirement* £	*Interest (1/12 × 18%)* £	*Actual requirement* £	*Interest (1/12 × 18%)* £
1	3 409	51	3 409	51
2	10 909	164	10 909	164
3	18 256	274	18 544	278
4	21 327	320	22 251	334
5	17 427	261	16 947	254
6	9 436	142	17 446	262
7	1 598	24	713	11
		1 236		1 354

Worksheet MII/90/5

The Chartered Institute of Building

BUILDING ECONOMICS AND FINANCE – WORKSHEET CASH FLOW ASSESSMENT CHART

Month	A	B	Actual Contract Expenditure	Actual money received (B – retention)	–ve shortfall	Actual monthly income (A – retention)	Gross
	Monthly Budget* £	Cumulative Budget* £	£	(B × 97%) £	Net cash flow £	(A × 97%) £	Max cash requirement £
1	(w1) 3 375	3 375	3 409	–	–3 409	–	–3 409
2	(w2) 7 425	10 800	10 909	3 274	–7 635	3 274	–10 909
3	13 200	24 000	21 818	10 476	–11 342	7 202	–18 544
4	(w3) 3 000	27 000	32 727	23 280	–9 447	12 804	–22 251
5	17 250	44 250	40 227	26 190	–14 037	2 910	–16 947
6	3 3750	48 000	43 636	42 923	+713	16 733	–17 446
7				46 560	+2 924	3 637	–713
				720	+720	720	
				47 280			
				720	+720	720	
				48 000		48 000	

*Adjusted for undervaluation

W1 90% × £3 750

W2 90% × £8 250

W3 25% × £12 000

Candidate number.............
MII/90/5/A

Question 5

BUILDING ECONOMICS AND FINANCE – INFORMATION AND WORKSHEET

Project value, cost and payment analysis

Month End	Cumulative value £	Cumulative cost £	Cumulative Payment £
May	81 000	73 636	–
June	232 000	210 909	–
July	358 000	325 455	76 950
August	554 000	503 636	220 400
September	634 000	576 364	340 100
October	700 000	636 364	526 300
November	–	–	602 300
December	–	–	682 500
June	–	–	700 000

Average payment delay statement

Input	Months 1–3		Months 4–6
Labour	10% × 1 week	= 0.10	25% × 1 week = 0.25
Materials	15% × 4 weeks	= 0.60	40% × 4 weeks = 1.60
Plant	60% × 5 weeks	= 3.00	15% × 5 weeks = 0.75
Sub-contractors	15% × 2 weeks	= 0.30	20% × 2 weeks = 0.40
	weighted delay	= 4.00	weighted delay = 3.00

(This sheet, when completed, must be handed in with the answer script).

The cumulative cost is 10/11 of the cumulative value for the same reasons as given in the previous solution.

Cumulative payment is time lagged and represents 95% of cumulative value.

The cumulative payment in December (£682,500) includes £17,500 being 50% of the retention, the remaining £17,500 is paid in June.

Candidate number. . . . .

MII/90/5/B

BUILDING ECONOMICS AND FINANCE – CASH FLOW WORKSHEET

Cash flow statement

Month	Cost		Income		Cash flow position	
	Beginning	End	Beginning	End	Beginning	End
May	–		–	–	–	
June		73 636	–	–		−73 636
July	73 636	210 909	76 950	76 950	−73 636	−133 959
August	210 909	325 455	220 400	220 400	−133 959	−105 055
Sept	325 455	503 636	340 100	340 100	−105 055	−163 536
Oct	503 636	576 364	526 300	526 300	−163 536	−50 064
Nov	576 364	636 364	602 300	602 300	−50 064	−34 064
Dec	636 364	636 364	682 500	682 500	−34 064	+ 46 136
June	636 364	636 364	700 000	–	+ 46 136	+63 636

(This sheet, when completed, must be handed in with the answer script).

The end cash flow position is the difference between the end columns of cost and income.

Chapter 17

Question 1

The calculation of standard cost labour variances is explained in Chapter 17.5 and illustrated in Figure 17.3

	£
Standard cost of actual production	
Standard hours × standard rate	
(360/4) × £4.80	432.00
Actual cost of actual production	
Actual hours × actual rate	
(2 × 39) × £5.00	390.00
Labour cost variance	42.00 (F)
Actual hours × standard rate	
(2 × 39) × £4.80	374.40
Actual hours × actual rate	
(2 × 39) × £5.00	390.00
Labour rate variance	15.60 (A)
Standard hours × standard rate	
(360/4) × £4.80	432.00
Actual hours × standard rate	
(2 × 39) × £4.80	374.40
Labour efficiency variance	57.60 (F)

Index